On the Shoulders of GIANTS

New Approaches to Numeracy

LYNN ARTHUR STEEN
Editor

Mathematical Sciences Education Board
National Research Council

NATIONAL ACADEMY PRESS
Washington, D.C. 1990

NATIONAL ACADEMY PRESS • 2101 Constitution Avenue, NW • Washington, DC 20418

NOTICE: The project that is the subject of this report was approved by the Governing Board of the National Research Council, whose members are drawn from the councils of the National Academy of Sciences, the National Academy of Engineering, and the Institute of Medicine. The authors of the report were chosen for their special competences.

The National Research Council was organized by the National Academy of Sciences in 1916 to associate the broad community of science and technology with the Academy's purposes of furthering knowledge and advising the federal government. Functioning in accordance with general policies determined by the Academy, the Council has become the principal operating agency of both the National Academy of Sciences and the National Academy of Engineering in providing services to the government, the public, and the scientific and engineering communities. The Council is administered jointly by both Academies and the Institute of Medicine. Dr. Frank Press and Dr. Robert M. White are chairman and vice chairman, respectively, of the National Research Council.

The Mathematical Sciences Education Board was established in 1985 to provide a continuing national overview and assessment capability for mathematics education and is concerned with mathematical sciences education for all students at all levels. The Board reports directly to the Governing Board of the National Research Council.

Development, publication, and dissemination of this book were supported by a grant from the Carnegie Corporation of New York. Additional dissemination of the book was supported by a grant from the Andrew W. Mellon Foundation. The observations made herein do not necessarily reflect the views of the grantors.

Library of Congress Cataloguing-in-Publication Data

On the shoulders of giants : new approaches to numeracy / Lynn Arthur Steen, editor ; Mathematical Sciences Education Board, National Research Council.
p. cm.
Includes bibliographical references and index.
ISBN 0-309-04234-8 (hardbound)
1. Mathematics—Study and teaching—United States. I. Steen, Lynn Arthur, 1941– II. National Research Council (U.S.). Mathematical Sciences Education Board.
QA13.053 1990
513'.071073—dc20 90-41566
CIP

Cover: A computer-generated image of a torus (the surface of a doughnut-shaped figure). When projected from four-dimensional space, the torus divides all of three-dimensional space into two symmetrical pieces. Created by Thomas Banchoff and Nicholas Thompson at Brown University.

Copies of this report may be purchased from the National Academy Press, 2101 Constitution Avenue, NW, Washington, DC 20418.

Printed in the United States of America

Preface

~ ~ ~ ~ ~ ~ ~ ~ ~ ~ ~ ~ ~ ~ ~ ~ ~ ~ ~ ~

Today's headlines are filled with reports of illiteracy, innumeracy, and other signs of educational decay. Tomorrow's schools can be filled with evidence of renaissance if we begin now to till the soil for effective education—in mathematics, in science, and in all subjects. This volume offers five visions of mathematics suitable for tomorrow's schools—visions rooted in imagination, in mathematics, and in science. Ideas in this volume can provide fertile soil for new approaches to tomorrow's numeracy.

Forces created by computers, applications, demographics, and schools themselves are changing profoundly the way mathematics is practiced, the way it is taught, and the way it is learned. Even as we work to make incremental change in today's schools, we must think also about more significant change that will be possible, indeed inevitable, in the future. For this reason the Mathematical Sciences Education Board (MSEB) decided that one of today's priorities is to stimulate imaginative thinking about tomorrow's curriculum.

In this volume readers will find a vision of the richness of mathematics expressed through five vignettes that illustrate different possible strands of school mathematics. These papers expand on the theme of mathematics as the language and science of patterns and are introduced by a brief essay that highlights interconnections and common ideas. The authors were asked to explore ideas with deep roots in the mathematical sciences without concern for limitations of present schools or curricula. They do, however, suggest through numerous imaginative examples how mathematical ideas can be developed from informal childhood exploration through formal school and college study.

The papers in this volume are intended as a vehicle to stimulate creative approaches to mathematics curricula in the next century. The volume itself is part of a national dialogue on mathematics education stimulated by a series of recent publications:

- *Everybody Counts: A Report to the Nation on the Future of Mathematics Education*
- *Curriculum and Evaluation Standards for School Mathematics*
- *Science for All Americans*
- *Reshaping School Mathematics: A Philosophy and Framework for Curriculum*

Taken together these publications provide a consistent and urgent vision that should help the United States restore excellence to mathematics education.

Although five examples are presented in this volume, they are certainly not the only five possibilities. Appropriate curricula for the twenty-first century will necessarily involve a wide variety of strands, reflecting both the broad spectrum of mathematical sciences and the individual choices of local school districts. We offer these themes not as definitive recommendations for curricula but as samples of what is possible, to stimulate development of new and imaginative programs that reflect the vitality and uses of mathematics.

Although each essay in this volume is the work of one author, each has benefited enormously from advice and critique provided by many advisers. Overall, the volume was developed under the auspices of the 1989 MSEB Curriculum Committee chaired by Henry O. Pollak, retired assistant vice president of Bell Communications Research. Other members of this Advisory Committee included Wade Ellis, Jr. of West Valley College; Andrew M. Gleason of Harvard University; Martin D. Kruskal of Princeton University; Leslie Paoletti of Choate Rosemary Hall; Anthony Ralston of the State University of New York at Buffalo; Isadore Singer of the Massachusetts Institute of Technology; and Zalman Usiskin of the University of Chicago. These individuals deserve much of the credit for helping shape the volume at its inception and for keeping it on track.

Seven "Sounding Boards" were established by the MSEB to review drafts of the essays as the volume progressed—one for the overview paper, one for each main essay, and one to examine the links with science. For "Pattern" the Sounding Board consisted of Isadore Singer and Zal Usiskin; for "Dimension," David Masunaga of the Iolani School in Honolulu and Jean Taylor of Rutgers University; for "Quantity," Harvey Keynes of the University of Minnesota and Alan Tucker of the State University of New York at Stony Brook; for "Uncertainty," James Landwehr of AT&T Bell Laboratories and James Swift of Nanaimo Senior Secondary School in British Columbia; for "Shape," Branko Grünbaum of the University of Washington and Paula Fitzmaurice of Victor J. Andrew High School in Tinley Park, Illinois; and for "Change," Robert Devaney of Boston University and Leslie Paoletti. The Scientific Sounding Board which reviewed the entire volume consisted of William O. Baker, retired chairman of Bell Laboratories; Maurice Fox, professor of biology at MIT; and Gerard Debreau, professor of economics at the University of California at Berkeley.

Many improvements in this volume are due directly to the hard work and good ideas of these distinguished Sounding Board reviewers. To be fair to them, however, it is important to acknowledge that the authors did not always heed the advice proffered by their reviewers; so while we are genuinely grateful for their assistance, full responsibility for the points of view expressed in this volume rests with the authors.

Publication of this volume completes the first phase in the work of the MSEB to express to the nation a new vision of mathematics education, of how a centuries-old curriculum can evolve to meet the challenges of the next millennium. Right from the beginning of MSEB in 1985, former MSEB Chair Shirley Hill of the University of Missouri at Kansas City took up the difficult challenge of forcing mathematicians and mathematics educators to think together about possible new strands for the mathematics curriculum. She challenged all of us on the MSEB to seek out ideas that may be more appropriate to our computer age than the arithmetic-bound structures that we have inherited from previous generations for whom calculation was the primary purpose of mathematics. This volume is the direct result of Shirley's persistence in emphasizing the importance of rooting curricular reform in the emerging practice of mathematics.

Coordination and production details have been ably managed by the MSEB staff led first by Marcia Sward and now by Kenneth Hoffman. Special thanks are due Linda Rosen, who shepherded with unfailing good humor all technical aspects of production from the initial planning meetings to final details of artwork, copy editing, and production. Thanks also are due Jana Godsey whose tenacity and patience were invaluable in collecting the many illustrations for the volume. Much of the computer-generated artwork was provided by Thomas Banchoff and David Moore, with special support from Davide Cervone, a graduate student at Brown University. Finally, throughout the many drafts of the different essays, Mary Kay Peterson managed with efficiency all the TEX typing and corrections necessary to enable the final text to be produced by direct electronic means.

Lynn Arthur Steen, Editor
St. Olaf College

Contents

Pattern

~ ~

LYNN ARTHUR STEEN

"He just saw further than the rest of us." The subject of this remark, cyberneticist Norbert Wiener, is one of many exceptional scientists who broke the bonds of tradition to create entirely new domains for mathematicians to explore. Seeing and revealing hidden patterns are what mathematicians do best. Each major discovery opens new areas rich with potential for further exploration. In the last century alone, the number of mathematical disciplines has grown at an exponential rate; examples include the ideas of Georg Cantor on transfinite sets, Sonja Kovalevsky on differential equations, Alan Turing on computability, Emmy Noether on abstract algebra, and, most recently, Benoit Mandelbrot on fractals.

To the public these new domains of mathematics are *terra incognita.* Mathematics, in the common lay view, is a static discipline based on formulas taught in the school subjects of arithmetic, geometry, algebra, and calculus. But outside public view, mathematics continues to grow at a rapid rate, spreading into new fields and spawning new applications. The guide to this growth is not calculation and formulas but an open-ended search for pattern.

Mathematics has traditionally been described as the science of number and shape. The school emphasis on arithmetic and geometry is deeply rooted in this centuries-old perspective. But as the territory explored by mathematicians has expanded—into group theory and statistics, into optimization and control theory—the historic boundaries of mathematics have all but disappeared. So have the boundaries of its applications: no longer just the language of physics and engineering, mathematics is now an essential tool for banking, manufacturing, social science, and medicine. When viewed in this broader context, we see that mathematics is not just about number and shape but about pattern and order of all sorts. Number and shape—arithmetic and geometry—are but two of many media in which mathematicians work. Active mathematicians seek patterns wherever they arise.

Thanks to computer graphics, much of the mathematician's search for patterns is now guided by what one can really see with the eye, whereas nineteenth-century mathematical giants like Gauss and Poincaré had to depend more on seeing with their mind's eye. "I see" has always had two distinct meanings: to perceive with the eye and to understand with the mind. For centuries the mind has dominated the eye in the hierarchy of mathematical practice; today the balance is being restored as mathematicians find new ways to see patterns, both with the eye and with the mind.

Change in the practice of mathematics forces re-examination of mathematics education. Not just computers, but also new applications and new theories have expanded significantly the role of mathematics in science, business, and technology. Students who will live and work using computers as a routine tool need to learn a different mathematics than their forefathers. Standard school practice, rooted in traditions that are several centuries old, simply cannot prepare students adequately for the mathematical needs of the twenty-first century.

Shortcomings in the present record of mathematical education also provide strong forces for change. Indeed, since new developments build on fundamental principles, it is plausible, as many observers often suggest, that one should focus first on restoring strength to time-honored fundamentals before embarking on reforms based on changes in the contemporary practice of mathematics. Public support for strong basic curricula reinforces the wisdom of the past—that traditional school mathematics, if carefully taught and well learned, provides sound preparation both for the world of work and for advanced study in mathematically based fields.

The key issue for mathematics education is not *whether* to teach fundamentals but *which* fundamentals to teach and *how* to teach them. Changes in the practice of mathematics do alter the balance of priorities among the many topics that are important for numeracy. Changes in society, in technology, in schools—among others—will have great impact on what will be possible in school mathematics in the next century. All of these changes will affect the fundamentals of school mathematics.

To develop effective new mathematics curricula, one must attempt to foresee the mathematical needs of tomorrow's students. It is the present and future practice of mathematics—at work, in science, in research—that should shape education in mathematics. To prepare effective mathematics curricula for the future, we must look to patterns in the mathematics of today to project, as best we can, just what is and what is not truly fundamental.

FUNDAMENTAL MATHEMATICS

School tradition has it that arithmetic, measurement, algebra, and a smattering of geometry represent the fundamentals of mathematics. But there is much more to the root system of mathematics—deep ideas that nourish the growing branches of mathematics. One can think of specific mathematical structures:

- Numbers
- Algorithms
- Ratios
- Shapes
- Functions
- Data

or attributes:

- Linear
- Periodic
- Symmetric
- Continuous
- Random
- Maximum
- Approximate
- Smooth

or actions:

- Represent
- Control
- Prove
- Discover
- Apply
- Model
- Experiment
- Classify
- Visualize
- Compute

or abstractions:

- Symbols
- Infinity
- Optimization
- Logic
- Equivalence
- Change
- Similarity
- Recursion

or attitudes:

- Wonder
- Meaning
- Beauty
- Reality

or behaviors:

- Motion
- Chaos
- Resonance
- Iteration
- Stability
- Convergence
- Bifurcation
- Oscillation

or dichotomies:

- Discrete vs. continuous
- Finite vs. infinite
- Algorithmic vs. existential
- Stochastic vs. deterministic
- Exact vs. approximate

These diverse perspectives illustrate the complexity of structures that support mathematics. From each perspective one can identify various strands that have within them the power to develop a significant mathematical idea from informal intuitions of early childhood all the way through school and college and on into scientific or mathematical research. A sound education in the mathematical sciences requires encounter with virtually all of these very different perspectives and ideas.

Traditional school mathematics picks very few strands (e.g., arithmetic, geometry, algebra) and arranges them horizontally to form the curriculum: first arithmetic, then simple algebra, then geometry, then more algebra, and finally—as if it were the epitome of mathematical knowledge—calculus. This layer-cake approach to mathematics education effectively prevents informal development of intuition along the multiple roots of mathematics. Moreover, it reinforces the tendency to design each course primarily to meet the prerequisites of the next course, making the study of mathematics largely an exercise in delayed gratification. To help students see clearly into their own mathematical futures, we need to construct curricula with greater vertical continuity, to connect the roots of mathematics to the branches of mathematics in the educational experience of children.

School mathematics is often viewed as a pipeline for human resources that flows from childhood experiences to scientific careers. The layers in the mathematics curriculum correspond to increasingly constricted sections of pipe through which all students must pass if they are to progress in their mathematical and scientific education. Any impediment to learning, of which there are many, restricts the flow in the entire pipeline. Like cholesterol in the blood, mathematics can clog the educational arteries of the nation.

In contrast, if mathematics curricula featured multiple parallel strands, each grounded in appropriate childhood experiences, the flow of human resources would more resemble the movement of nutrients in the roots of a mighty tree—or the rushing flow of water from a vast watershed—than the increasingly constricted confines of a narrowing artery or pipeline. Different aspects of mathematical experience will attract children of different interests and talents, each nurtured by challenging ideas that stimulate imagination and promote exploration. The collective effect will be to develop among children diverse mathematical insight in many different roots of mathematics.

FIVE SAMPLES

This volume offers five examples of the developmental power of deep mathematical ideas: dimension, quantity, uncertainty, shape, and change. Each chapter explores a rich variety of patterns that can be introduced to children at various stages of school, especially at the youngest ages when unfettered curiosity remains high. Those who develop curricula will find in these essays many valuable new options for school mathematics. Those who help determine education policy will see in these essays examples of new standards for excellence. And everyone who is a parent will find in these essays numerous examples of important and effective mathematics that could excite the imagination of their children.

Each chapter is written by a distinguished scholar who explains in everyday language how fundamental ideas with deep roots in the mathematical sciences could blossom in schools of the future. Although not constrained by particular details of present curricula, each essay is faithful to the development of mathematical ideas from childhood to adulthood. In expressing these very different strands of mathematical thought, the authors illustrate ideals of how mathematical ideas should be developed in children.

In contrast to much present school mathematics, these strands are alive with action: pouring water to compare volumes, playing with pendulums to explore dynamics, counting candy colors to grasp variation, building kaleidoscopes to explore symmetry. Much mathematics can be learned informally by such activities long before children reach the point of understanding algebraic formulas. Early experiences with such patterns as volume, similarity, size, and randomness prepare students both for scientific investigations and for more formal and logically precise mathematics. Then when a careful demonstration emerges in class some years later, a student who has benefited from substantial early informal mathematical experiences can say with honest pleasure "Now I see why that's true."

CONNECTIONS

The essays in this volume are written by five different authors on five distinct topics. Despite differences in topic, style, and approach, these essays have in common the lineage of mathematics: each is connected in myriad ways to the family of mathematical sciences. Thus it should come as no surprise that the essays themselves are replete with interconnections, both in deep structure and even in particular illustrations. Some examples:

MEASUREMENT is an idea treated repeatedly in these essays. Experience with geometric quantities (length, area, volume), with arithmetic quantities (size, order, labels), with random variation (spinners, coin tosses, SAT scores), and with dynamic variables (discrete, continuous, chaotic) all pose special challenges to answer a very child-like question: "How big is it?" One sees from many examples that this question is fundamental: it is at once simple yet subtle, elementary yet difficult. Students who grow up recognizing the complexity of measurement may be less likely to accept unquestioningly many of the common misuses of numbers and statistics. Learning how to measure is the beginning of numeracy.

SYMMETRY is another deep idea of mathematics that turns up over and over again, both in these essays and in all parts of mathematics. Sometimes it is the symmetry of the whole, such as the hypercube (a four-dimensional cube), whose symmetries are so numerous that it is hard to count them all. (But with proper guidance, young children using a simple pea-and-toothpick model can do it.) Other times it is the symmetry of the parts, as in the growth of natural objects from repetitive patterns of molecules or cells. In still other cases it is symmetry broken, as in the buckling of a cylindrical beam or the growth of a fertilized egg to a (slightly) asymmetrical adult animal. Unlike measurement, symmetry is seldom studied much in school at any level, yet it is equally fundamental as a model for explaining features of such diverse phenomena as the basic forces of nature, the structure of crystals, and the growth of organisms. Learning to recognize symmetry trains the mathematical eye.

VISUALIZATION recurs in many examples in this volume and is one of the most rapidly growing areas of mathematical and scientific research. The first step in data analysis is the visual display of data to search for hidden patterns. Graphs of various types provide visual display of relations and functions; they are widely used throughout science and industry to portray the behavior of one variable (e.g., sales) that is a function of another (e.g., advertising). For centuries artists and map makers have used geometric devices such as projection to represent three-dimensional scenes on a two-dimensional canvas or sheet of paper. Now computer graphics automate these processes and let us explore as well the projections of shapes in higher-dimensional space. Learning to visualize mathematical patterns enlists the gift of sight as an invaluable ally in mathematical education.

ALGORITHMS are recipes for computation that occur in every corner of mathematics. A common iterative procedure for projecting population growth reveals how simple orderly events can lead to a variety of behaviors—explosion, decay, repetition, chaos. Exploration of combinatorial patterns in geometric forms enables students to project geometric structures in higher dimensions where they cannot build real models. Even common elementary school algorithms for arithmetic take on a new dimension when viewed from the perspective of contemporary mathematics: rather than stressing the mastery of specific algorithms—which are now carried out principally by calculators or computers—school mathematics can instead emphasize more fundamental attributes of algorithms (e.g., speed, efficiency, sensitivity) that are essential for intelligent use of mathematics in the computer age. Learning to think algorithmically builds contemporary mathematical literacy.

Many other connective themes recur in this volume, including linkages of mathematics with science, classification as a tool for understanding, inference from axioms and data, and—most importantly—the role of exploration in the process of learning mathematics. Connections give mathematics power and help determine what is fundamental. Pedagogically, connections permit insight developed in one strand to infuse into others. Multiple strands linked by strong interconnections can develop mathematical power in students with a wide variety of enthusiasms and abilities.

GAINING PERSPECTIVE

Newton credited his extraordinary foresight in the development of calculus to the accumulated work of his predecessors: "If I have seen farther than others, it is because I have stood on the shoulders of giants." Those who develop mathematics curricula for the twenty-first century will need similar foresight.

Not since the time of Newton has mathematics changed as much as it has in recent years. Motivated in large part by the introduction of computers, the nature and practice of mathematics have been fundamentally transformed by new concepts, tools, applications, and methods. Like the telescope of Galileo's era that enabled the Newtonian revolution, today's computer challenges traditional views and forces re-examination of deeply held values. As it did three centuries ago in the transition from Euclidean proofs to Newtonian analysis, mathematics once again is undergoing a fundamental reorientation of procedural paradigms.

Examples of fundamental change abound in the research literature of mathematics and in practical applications of mathematical methods. Many are given in the essays in this volume:

- Uncertainty is not haphazard, since regularity eventually emerges.
- Deterministic phenomena often exhibit random behavior.
- Dimensionality is not just a property of space but also a means of ordering knowledge.
- Repetition can be the source of accuracy, symmetry, or chaos.
- Visual representation yields insights that often remain hidden from strictly analytic approaches.
- Diverse patterns of change exhibit significant underlying regularity.

By examining many different strands of mathematics, we gain perspective on common features and dominant ideas. Recurring concepts (e.g., number, function, algorithm) call attention to what one must know in order to *understand* mathematics; common actions (e.g., represent, discover, prove) reveal skills that one must develop in order to *do* mathematics. Together, concepts and actions are the nouns and verbs of the language of mathematics.

What humans do with the language of mathematics is to describe patterns. Mathematics is an exploratory science that seeks to understand every kind of pattern—patterns that occur in nature, patterns invented by the human mind, and even patterns created by other patterns. To grow mathematically, children must be exposed to a rich variety of patterns appropriate to their own lives through which they can see variety, regularity, and interconnections.

The essays in this volume provide five extended case studies that exemplify how this can be done. Other authors could just as easily have described five or ten different examples. The books and articles listed below are replete with additional examples of rich mathematical ideas. What matters in the study of mathematics is not so much which particular strands one explores, but the presence in these strands of significant examples of sufficient variety and depth to reveal patterns. By encouraging students to explore patterns that have proven their power and significance, we offer them broad shoulders from which they will see farther than we can.

REFERENCES AND RECOMMENDED READING

1. Albers, Donald J. and Alexanderson, G.L. *Mathematical People: Profiles and Interviews.* Cambridge, MA: Birkhauser Boston, 1985.
2. Barnsley, Michael F. *Fractals Everywhere.* New York, NY: Academic Press, 1988.
3. Barnsley, Michael F. *The Desktop Fractal Design System.* New York, NY: Academic Press, 1989.
4. Brook, Richard J., et al. (Eds.). *The Fascination of Statistics.* New York, NY: Marcel Dekker, 1986.

5. Campbell, Stephen K. *Flaws and Fallacies in Statistical Thinking.* Englewood Cliffs, NJ: Prentice-Hall, 1974.
6. Davis, Philip J. and Hersh, Reuben. *Descartes' Dream: The World According to Mathematics.* San Diego, CA: Harcourt Brace Jovanovich, 1986.
7. Davis, Philip J. and Hersh, Reuben. *The Mathematical Experience.* Boston, MA: Birkhauser, 1980.
8. Devaney, Robert L. *Chaos, Fractals, and Dynamics: Computer Experiments in Mathematics.* Reading, MA: Addison-Wesley, 1990.
9. Dewdney, A.K. *The Turing Omnibus: 61 Excursions in Computer Science.* Rockville, MD: Computer Science Press, 1989.
10. Ekeland, Ivar. *Mathematics and the Unexpected.* Chicago, IL: University of Chicago Press, 1988.
11. Fischer, Gerd. *Mathematical Models from the Collections of Universities and Museums.* Wiesbaden, FRG: Friedrich Vieweg & Sohn, 1986.
12. Francis, George K. *A Topological Picturebook.* New York, NY: Springer-Verlag, 1987.
13. Gleick, James. *Chaos.* New York, NY: Viking Press, 1988.
14. Guillen, Michael. *Bridges to Infinity: The Human Side of Mathematics.* Boston, MA: Houghton Mifflin, 1983.
15. Hoffman, Paul. *Archimedes' Revenge: The Joys and Perils of Mathematics.* New York, NY: W.W. Norton & Company, 1988.
16. Hofstadter, Douglas R. *Gödel, Escher, Bach: An Eternal Golden Braid.* New York, NY: Vintage Press, 1980.
17. Holden, Alan. *Shapes, Space, and Symmetry.* New York, NY: Columbia University Press, 1971.
18. Huff, Darrell. *How to Lie with Statistics.* New York, NY: W.W. Norton & Company, 1954.
19. Huntley, H.E. *The Divine Proportion: A Study in Mathematical Beauty.* Mineola, NY: Dover, 1970.
20. Jaffe, Arthur. "Ordering the Universe: The Role of Mathematics." In *Renewing U.S. Mathematics: Critical Resource for the Future.* Washington, DC: National Academy Press, 1984, 117–162.
21. Kitcher, Philip. *The Nature of Mathematical Knowledge.* New York, NY: Oxford University Press, 1983.
22. Kline, Morris. *Mathematics and the Search for Knowledge.* New York, NY: Oxford University Press, 1985.
23. Lang, Serge. *MATH! Encounters with High School Students.* New York, NY: Springer-Verlag, 1985.
24. Lang, Serge. *The Beauty of Doing Mathematics: Three Public Dialogues.* New York, NY: Springer-Verlag, 1985.
25. Loeb, Arthur. *Space Structures: Their Harmony and Counterpoint.* Reading, MA: Addison-Wesley, 1976.
26. Mandelbrot, Benoit B. *The Fractal Geometry of Nature.* New York, NY: W.H. Freeman, 1982.
27. Moore, David S. *Statistics: Concepts and Controversies, Second Edition.* New York, NY: W.H. Freeman, 1985.
28. Morrison, Philip and Morrison, Phylis. *Powers of Ten.* New York, NY: Scientific American Books, 1982.
29. Peitgen, Heinz-Otto and Richter, Peter H. *The Beauty of Fractals: Images of Complex Dynamical Systems.* New York, NY: Springer-Verlag, 1986.
30. Peitgen, Heinz-Otto and Saupe, Dietmar (Eds.). *The Science of Fractal Images.* New York, NY: Springer-Verlag, 1988.

31. Peterson, Ivars. *The Mathematical Tourist: Snapshots of Modern Mathematics.* New York, NY: W.H. Freeman, 1988.
32. Rosen, Joe. *Symmetry Discovered: Concepts and Applications in Nature and Science.* New York, NY: Cambridge University Press, 1975.
33. Rucker, Rudy. *Infinity and the Mind: The Science and Philosophy of the Infinite.* Boston, MA: Birkhauser, 1982.
34. Rucker, Rudy. *The Fourth Dimension: Toward a Geometry of Higher Reality.* Boston, MA: Houghton Mifflin, 1984.
35. Senechal, Marjorie and Fleck, George (Eds.). *Patterns of Symmetry.* Amherst, MA: University of Massachusetts Press, 1977.
36. Sondheimer, Ernst and Rogerson, Alan. *Numbers and Infinity: A Historical Account of Mathematical Concepts.* New York, NY: Cambridge University Press, 1981.
37. Steen, Lynn Arthur. *Mathematics Today: Twelve Informal Essays.* New York, NY: Springer-Verlag, 1978.
38. Steen, Lynn Arthur. "The Science of Patterns." *Science,* 240 (29 April 1988), 611–616.
39. Steinhaus, H. *Mathematical Snapshots, Third American Edition Revised and Enlarged.* New York, NY: Oxford University Press, 1983.
40. Stevens, Peter S. *Patterns in Nature.* Boston, MA: Little, Brown & Company, 1974.
41. Stewart, Ian. *The Problems of Mathematics.* New York, NY: Oxford University Press, 1987.
42. Stewart, Ian. *Does God Play Dice? The Mathematics of Chaos.* Oxford: Blackwell, 1989.
43. Tanur, Judith M., et al. (Eds.). *Statistics: A Guide to the Unknown, Third Edition.* Laguna Hills, CA: Wadsworth, 1989.
44. Tufte, Edward R. *The Visual Display of Quantitative Information.* Cheshire, CT: Graphics Press, 1983.
45. Wenninger, Magnus J. *Polyhedron Models for the Classroom, Second Edition.* Reston, VA: National Council of Teachers of Mathematics, 1975.

Dimension

~ ~

THOMAS F. BANCHOFF

INTRODUCTION

One hundred and fifty years ago, Friedrich Froebel (Figure 1), the inventor of the term "kindergarten," devised a set of "gifts" to introduce children to notions of geometry in several different dimensions. His philosophy was clear: if children could be stimulated to observe geometric objects from the earliest stage of their education, these ideas would come back to them again and again during the course of their schooling, deepening with each new level of sophistication. The rudimentary appreciation of shapes and forms at the nursery school level would become more refined as students developed new skills in arithmetic and measurement and later in more formal algebra and geometry.

In order to capture the imaginations of his young students, Froebel presented them with a sequence of wooden objects for their play in the Children's Garden. Only later would the lessons of that directed set of play experiences be turned into concepts and even later formalized into mathematical expressions. The important thing was to introduce students to forms that they could apprehend and to encourage them to observe and recognize those forms in all of their experiences. In this way they could foster the facility of visualization, so important in applying mathematics to both scientific and artistic pursuits.

Froebel began with objects from the most concrete part of mathematics: balls, cubes, and cylinders. He proceeded to a higher level of abstraction by presenting the children with trays covered by patterns of

FIGURE 1. Friedrich Froebel, inventor of kindergarten, used geometric objects to stimulate children's imaginations.

tiles. Then he moved further into abstraction by introducing collections of sticks of varying lengths, to be placed in designs that would ultimately be related to number patterns.

We can recognize some of Froebel's legacy in materials that we find in today's kindergarten classrooms. There we still have blocks for stacking and tiles for creating patterns on tabletops. Too often, however, these "toys" are left behind when children progress into the serious world of elementary school. A great many rods are used for arithmetic exercises, but a student is lucky to see anything two-dimensional between kindergarten and junior high school. At that time there might be a brief mention of area of plane figures, often merely as an illustration of formulas for measurement. Then the student must wait until high school before any further thought is given to the world of plane geometry.

Two generations ago the hardy souls who made it through the year of formal geometry were permitted to re-enter the third dimension in a still more formalized semester of solid geometry. Then curricula changed.

Three-dimensional topics (along with all of analytic geometry) were supposed to be incorporated into a single geometry curriculum. All too often the solid geometry components were treated merely as supplementary topics for the interested student who had a bit of leisure time. Needless to say, solid geometry quickly evaporated from the standard course in geometry. In the present-day rush to prepare students for calculus before they go off to college, we are systematically shortchanging them by ignoring the most practical and useful of all geometry—the geometry of our own dimension. We now have a special opportunity to bring the appreciation of different dimensions back into focus.

The Dimensional Ladder

Although our world is three-dimensional, most of our media, as it happens, are two-dimensional: blackboards, books, movies, television, and computer screens. We all invest a great deal of effort learning how to interpret such planar visual information, often in order to help us deal with situations in three dimensions. To live in a three-dimensional world, we do have to know how two-dimensional shapes interact: their behavior provides a necessary prelude to understand fully our own dimension.

As it happens, we gain a good deal of insight by investigating the geometry of an even lower dimension—the line—where number and geometry intermix in the most intimate and powerful way. The geometry of the number line translates beautifully into plane geometry, both in its classical form and in the analytic geometry of number pairs. The momentum that we gain in moving from the first to the second dimension can carry us into our home dimension with renewed insight. The dimensional analogy is a very powerful tool.

Here is an exciting theme that is worth recognizing and passing on to our students: the momentum that brings us from one to two and up into three dimensions does not stop there! The invitation is clear: there are other dimensions waiting to be explored. Mathematics is the key to the elevator that makes them accessible.

The fourth dimension, in particular, is one of our nearest neighbors. Just as we learn a good deal about our own language and culture by studying the language and culture of other countries, so we can begin to appreciate new things about our own "real" world by seeing structures that carry forward to the fourth dimension. Although we cannot explore higher dimensions physically, they are accessible to our minds and, thanks to modern technology, more and more to our vision as well.

Research into language acquisition indicates that, although any infant is capable of learning any language, a child will rather quickly settle into the sound patterns of its own particular language, effectively blocking development of other possibilities. If a child is not introduced early to other languages, he or she will experience much more difficulty in learning a second tongue. Might the same be true with respect to mathematical perceptions? If we wait until students have developed a great deal of arithmetic sophistication (and a great many misconceptions) before we encourage them to think about solid objects and the interaction between different dimensions, we may be depriving them of the chance to appreciate the full power and scope of geometry.

Giving Geometrical Gifts

Objects should always be nearby. Awareness of space and volume should be a continuing part of mathematical experience in school at all levels. Refinements such as measuring quantities and relating them with formulas will come in good time. But they should come well after the time when a child first becomes aware of different dimensions of measurement. Too often, the first time a student is encouraged to think about what volume means is the same day that he or she is given a formula for the volume of a sphere or a cone. To encourage fluency in the language of geometry, we need a good deal more "pre-geometry" throughout the school experience, and that should include "pre-solid" as well as "pre-plane" geometry.

Froebel and his colleagues created geometrical gifts from materials available to them, primarily wood, paper, and clay. Today we have the means to improve on the gifts in many ways—with plastic and Velcro, with tape and magnets, not to mention with the powerful computer graphics. The educator's term "manipulatives"—classroom materials—takes on new meaning when we can put in front of a young student a tool to manipulate not only simple forms but also the very geometry of higher dimensional space. If we care about educating our children toward the perception of space, we should create truly stimulating manipulatives—geometrical gifts for our day.

MEASURING VOLUMES

Many students never learn about volumes because they do not make it past plane geometry. Those who do often reach calculus by a head-long rush that leaves little or no time for the kind of geometrical thinking on which calculus thrives. Calculus is not the time when students should be doing their first serious thinking about geometry. Rather it should be

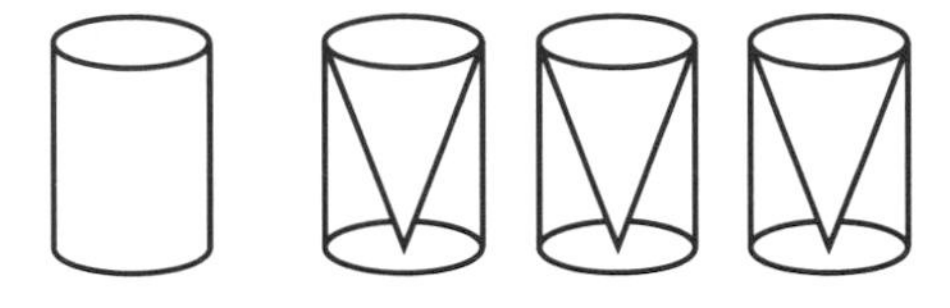

FIGURE 2. Water in a cylinder exactly fills three cones whose base and height are identical to the base and height of the cylinder.

the culmination of years of consideration of increasingly sophisticated geometrical topics. When a student finally sees the full justification of the formula for the volume of a cone or a sphere, it should be a peak experience, fulfilling a promise implicit in all the experiences he or she has had with cones and spheres all the way through school, beginning in kindergarten.

Froebel's young students spent a great deal of time pouring water and sifting sand. Differently shaped containers held different amounts, so a student would gradually learn common relationships without even thinking of writing them down. For example, how many conical cups can be filled from the water in a cylindrical cup with the same height and the same base? With a rack of such cups (Figure 2), any student can perform the experiment. The cylinder fills three cups.

We can test this over and over again with different heights and different circular tops. Only later, after the student is familiar with the language of fractions, need this relationship be stated in terms of one volume being one-third of another. Still later, that relationship can be expressed by a formula: the volume of the cone is one-third the area of the base multiplied by the height.

By this time that relationship should already have been observed in other shapes. Three square-based pyramids can be filled with the sand from one square prism of the same base and height (Figure 3). Even if the base is irregular, this relationship is true. We don't even have to have the center of the cone over the center of the base, assuming that the base even has a center! All this understanding can take place before the student has even seen a fraction, let alone a number like π.

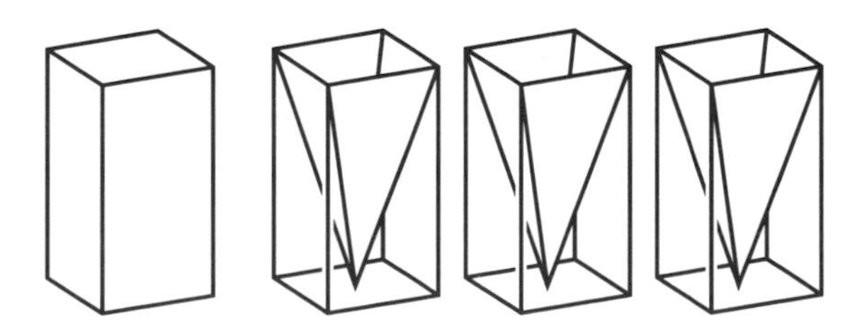

FIGURE 3. Water in a prism exactly fills three pyramids whose base and height match those of the prism. This relation holds even for prisms and cones with irregular bases and can be discovered by young children just by pouring water or sand.

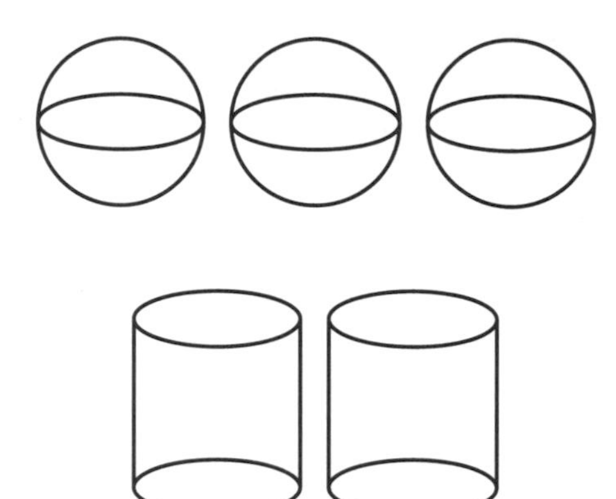

FIGURE 4. Pouring water can also verify Archimedes' theorem: the volume of three spheres equals the volume of two cylinders whose radius and height match those of the spheres.

A bit more subtle and even more impressive is the relationship that was symbolized on the gravestone of Archimedes: if a ball fits precisely inside a circular cylinder, then the volume of the ball is two-thirds the volume of the cylinder. To illustrate this we can show that three spheres can be filled with the water from two cylinders that encase the spheres (Figure 4). Volumes of irregularly shaped objects can be found by seeing how much water they displace when they are completely submerged. This leads naturally to the notion of density, as a weight-to-volume ratio.

The notion of area can be introduced by working with volumes. By using a collection of shallow pans, all of the same height, children can compare their volumes and relate them to the areas of their bases. The height dimension is "washed out" if it is the same in all cases. In this way it is easy for children to see that the area of a right triangle is half the area of the associated rectangle and that the area of a scalene triangle is half the area of three different associated parallelograms (Figure 5).

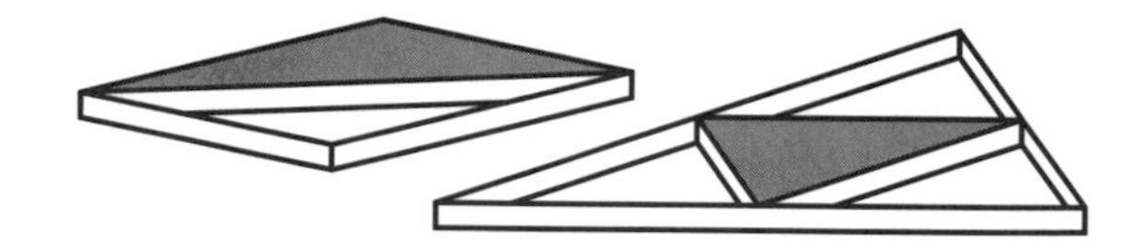

FIGURE 5. By pouring water into shallow pans, children can readily compare the areas of different geometric figures.

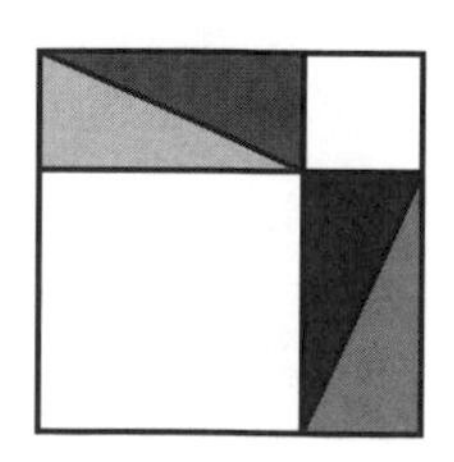

FIGURE 6. Four right triangles in a square frame reveal a proof of the Pythagorean theorem: the square on the hypotenuse equals the sum of the squares on the legs of the right triangle.

We can work as well with tiles of uniform thickness, as Froebel did in his kindergarten gifts in the last century. The relation between the area of a parallelogram and the area of a rectangle can be appreciated at a very early stage by students who actually manipulate physical objects. It isn't necessary to wait until students have learned about square roots before they can see an illustration of the Pythagorean theorem (Figure 6). Children who play with geometric puzzles that illustrate decompositions will find it much easier later on to appreciate formal results.

Decomposition Models

One of the most beautiful results that can be illustrated by blocks is the fact that a cube can be decomposed into three identical pieces meeting along a diagonal of the cube (Figure 8), just as a square is decomposed into two congruent triangles by a diagonal line (Figure 7).

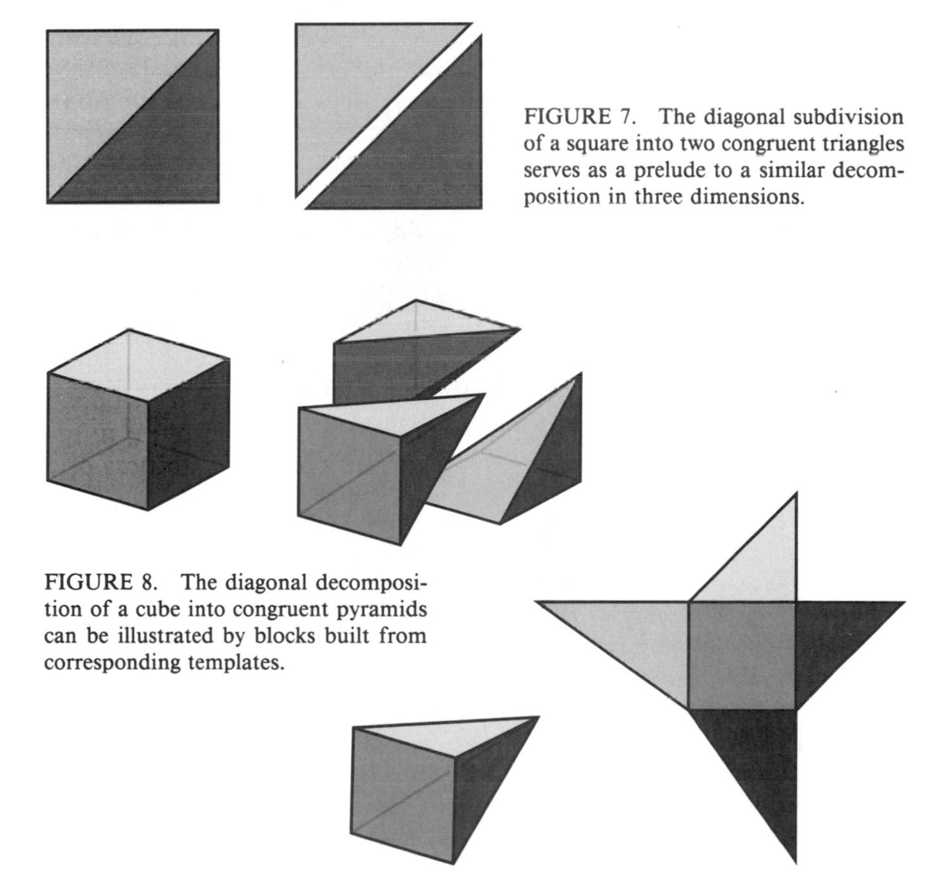

FIGURE 7. The diagonal subdivision of a square into two congruent triangles serves as a prelude to a similar decomposition in three dimensions.

FIGURE 8. The diagonal decomposition of a cube into congruent pyramids can be illustrated by blocks built from corresponding templates.

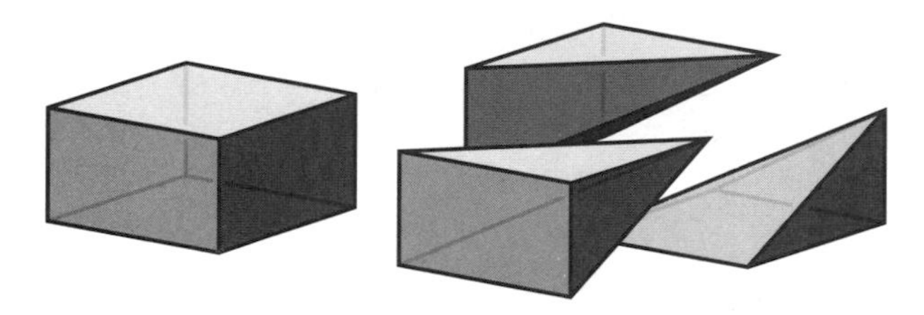

FIGURE 9. The diagonal decomposition of a rectangular solid yields three pyramids of different shapes but the same volume.

Decomposition models illustrate deeper ideas than do comparisons of volumes since they not only demonstrate relationships but also show *why* these relationships hold. Students should eventually come to see that all geometric relationships are based on reasons.

This particular decomposition property of the cube can be a bit misleading because it doesn't quite work for other rectangular solids. Although a diagonal always decomposes a rectangle into two congruent triangles, the diagonal decomposition of a rectangular solid will usually not produce three congruent pyramids (Figure 9). The three pyramidal parts will all have the same volume but not the same shape. This can be seen by pouring sand into plastic pyramid containers, but greater insight comes from a different model—playing cards.

Think of a pyramid constructed of thick rectangular cards stacked above the base. If we double the thickness of each card in the stack, then the base stays the same while both the height and the weight of the stack (and therefore its volume) also double. If we keep the width and the thickness of each card the same and double the length, then the volume also doubles. Doubling any single dimension causes the volume to double: in general, multiplying a single dimension by any number will multiply the entire volume by that same number.

This procedure enables us to obtain the volume of any pyramid formed by a diagonal decomposition of a rectangular solid—that is, of any pyramid with a rectangular base whose top vertex is directly over a corner of the base. Further work with pyramid-shaped blocks will quickly show that any pyramid with a rectangular base can be built up from pyramids of this special type, all with the same height. Taken together, these demonstrations show why, in general, the volume of a pyramid with a rectangular base is one-third the volume of the right rectangular prism with the same base and height.

Experiments with stacks of cards or thin rods can lead easily to a powerful idea known to mathematicians as Cavalieri's principle for shear transformations. First observe how the same set of rods that fills a parallelogram will also fill a rectangle with the same base and height. Hence

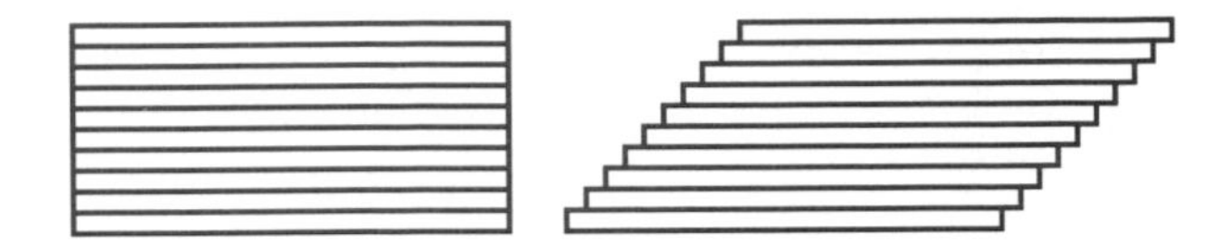

FIGURE 10. The same set of rods that forms a rectangle can also form a parallelogram of the same dimensions. Hence the areas of the rectangle and the parallelogram must be the same.

their areas must be equal (Figure 10). The same principle works in space as well as in the plane. The same pack of cards that fills a straight box can fill a slanted box with the same base and height. Similarly, an off-center pyramid can be approximated with the same collection of square cards that approximate a centered one (Figure 11).

Students who explore models of pyramids with sets of blocks and stacks of cards throughout their early school years are certainly more likely to understand and appreciate the formal proofs presented for such theorems in calculus classes; students who have never thought about properties of volumes until they arise in calculus will not get nearly as much out of their experience. We now spend a great deal of effort getting

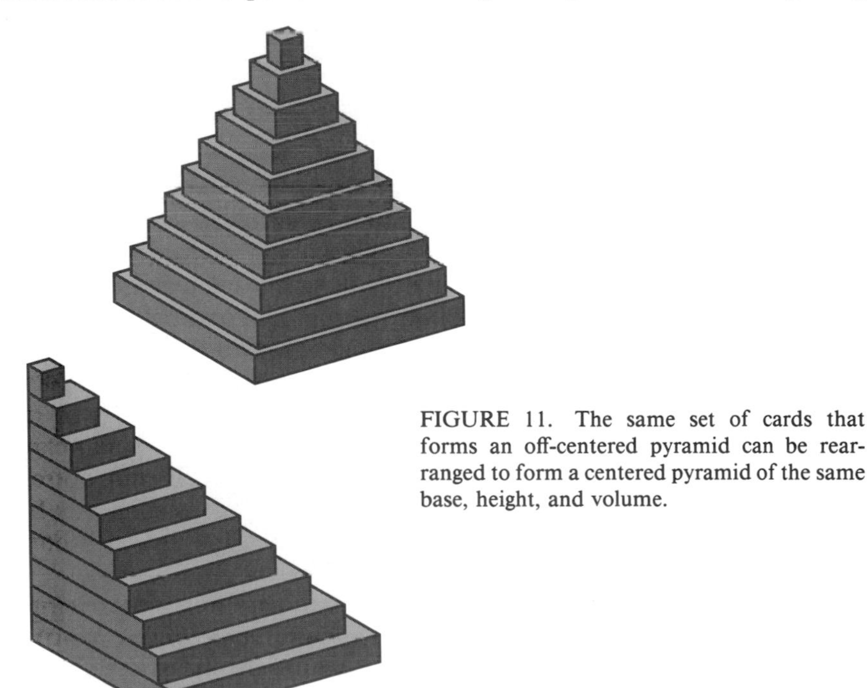

FIGURE 11. The same set of cards that forms an off-centered pyramid can be rearranged to form a centered pyramid of the same base, height, and volume.

students ready for the algebraic techniques needed for advanced mathematics. We should be just as concerned for their geometric preparation as well.

Pyramid Problems

Many children are fascinated by the great pyramids of Cheops. These only surviving wonders of the ancient world were mathematical challenges to their creators, and they remain challenging today. School study of the monuments of ancient Egypt can be a source of mathematics problems of all sorts, from the most elementary considerations of shadows to the most sophisticated achievement of early mensuration—the volume formula for the frustrum of a truncated pyramid.

Children can decide how to make models of the pyramids. A pile of dry sand or wet sand on a square base provides one example. Models in clay provide another. Students can experiment with different sizes of triangles to see what shapes of pyramids result.

Other monuments of different shapes provide similar exercises in measurement and challenges for construction. What about the burial mounds of American Indians or other cone-shaped structures? What about Mayan pyramids, with their step-like structure? What about Babylonian ziggerats, or pagodas? Each structure provides distinctive features that lead to interesting mathematical questions, which the students themselves can formulate and explore.

A key mathematical notion that arises naturally in the study of monuments is similarity, expressed both algebraically in ratio or proportion and geometrically in shadows and scale diagrams. Consider the following story:

> My friend Ambrose sent a snapshot of his trip to Egypt. He is standing next to an obelisk and I can see that his shadow is about one-fourth as long as the shadow of the obelisk. That's a pretty big column, over 24 feet high. I know that because my friend is 6 feet tall. There is a pyramid in the picture too. I can see that its shadow is falling just past the edge of the base. What additional information would I need in order to figure out how high the pyramid is? How can I measure the angle that the slanting side of the pyramid makes with the ground?

Such questions can be discussed at an informal level long before the students deal with triangles formally in geometry and trigonometry.

Thinking about the pyramids can show how problems in different dimensions can illuminate each other. Using the principle of similarity, students can easily calculate the volume of an incomplete pyramid (Figure 12), one of the most important problems in Egyptian mathematics.

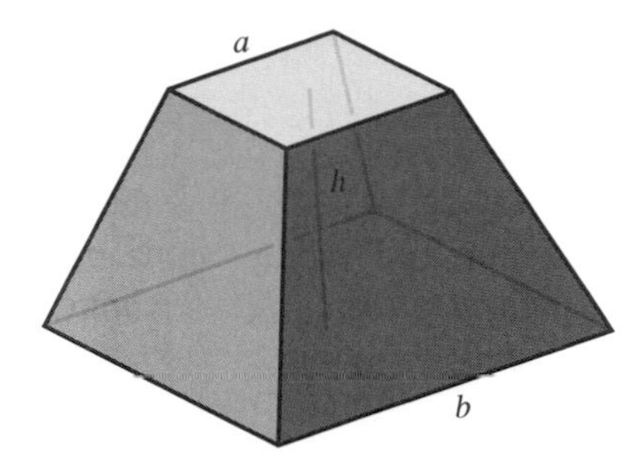

FIGURE 12. An incomplete (or truncated) pyramid poses a challenge to find its volume.

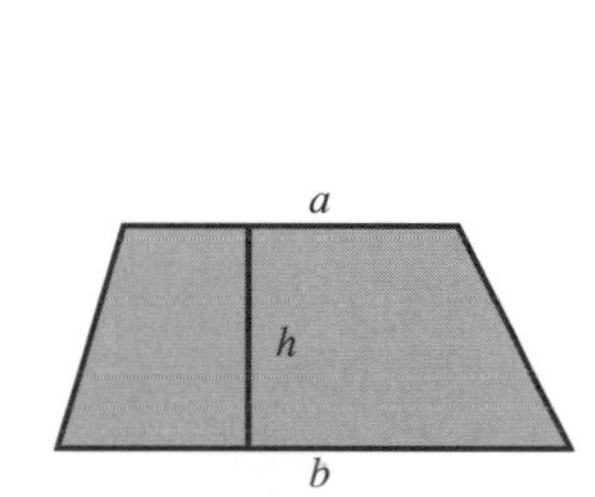

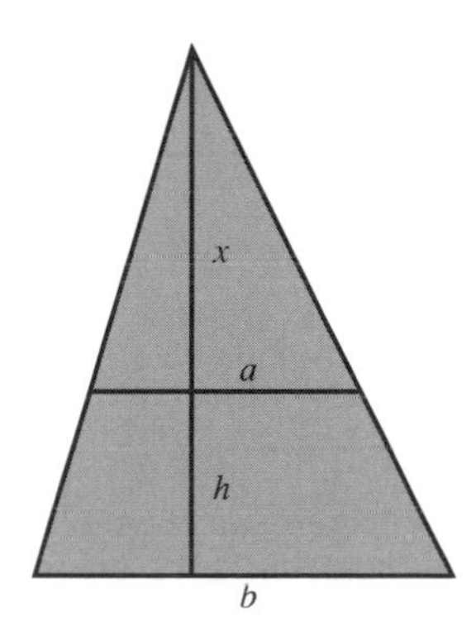

FIGURE 13. By thinking of a trapezoid as an incomplete triangle, we can find a way to calculate its area that can also be used in three dimensions to find the volume of an incomplete pyramid.

Begin with the analogous problem in the plane: the trapezoid viewed as an incomplete triangle (Figure 13). We know the quantities a, b, and h, and we want to find the area. Assuming that the trapezoid is not a parallelogram, we can complete the figure to a triangle with height that we call x. By observing that the large and small triangles are similar, we see that $x/a = (x+h)/b$. Hence $bx = ax + ah$, so $x = ha/(b-a)$ and $x + h = hb/(b-a)$. We then get the familiar formula for area of the trapezoid in a new way, as the difference of the areas of two triangles:

$$\begin{aligned} &(1/2)(x+h)b - (1/2)xa \\ &= (1/2)hb^2/(b-a) - (1/2)ha^2/(b-a) \\ &= (1/2)h(b^2 - a^2)/(b-a) \\ &= (1/2)h(b+a). \end{aligned}$$

The same method enables one to calculate the volume of the incomplete pyramid (Figure 14). We are given the height h of part of the pyramid and the side lengths a and b of the top and bottom squares. If the height of the large pyramid is $(x+h)$, then its total volume will be $(1/3)(x+h)b^2$, while the volume of the small pyramid is $(1/3)xa^2$.

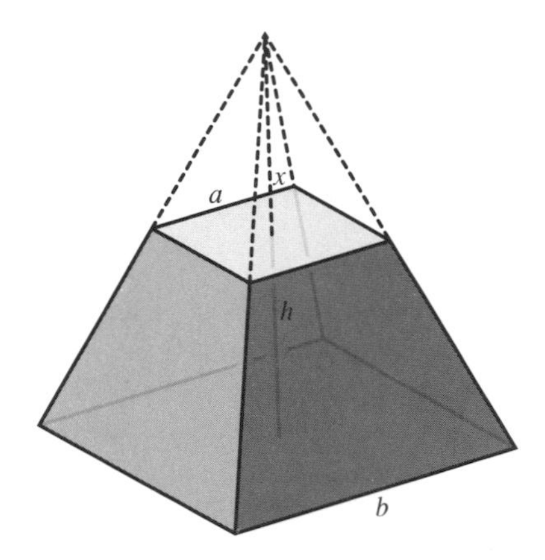

FIGURE 14. By completing the incomplete pyramid, its volume can be calculated as the difference of the volumes of two similar pyramids.

By similar triangles, $x/a = (x + h)/b$. So, as in the planar case, $x = ha/(b - a)$ and $x + h = hb/(b - a)$. Therefore the volume of the incomplete pyramid is

$$\begin{aligned} & (1/3)(x+h)b^2 - (1/3)xa^2 \\ &= (1/3)hb^3/(b-a) - (1/3)ha^3/(b-a) \\ &= (1/3)h(b^3 - a^3)/(b-a) \\ &= (1/3)h(b^2 + ab + a^2). \end{aligned}$$

This formula, which was detailed in a papyrus from 1800 B.C., represents a high point in the geometry of the ancient world. It can be appreciated by any student who reaches the level of first-year algebra. Truly enterprising students can conjecture the formula for the volume of an incomplete pyramid in the fourth dimension or in higher dimensions.

Cylinders and Discs

The volume of water in a circular cylinder is a little more than three-quarters of the volume of the rectangular box in which the cylinder just fits (Figure 15). If we pour the water from the cylinder into box-shaped containers of the same height, with square base whose side equals the radius of the cylinder, then we can fill three such boxes and still have some water left over. Experiments with different cylinders and related boxes will quickly show that this pattern works for cylinders of any radius or height. The same ratio, of course, relates the area of a circle to its circumscribing square. Because children can measure poured quantities more easily than painted areas, it may be easier for them to grasp this fundamental ratio first in terms of volume and then subsequently in terms of area.

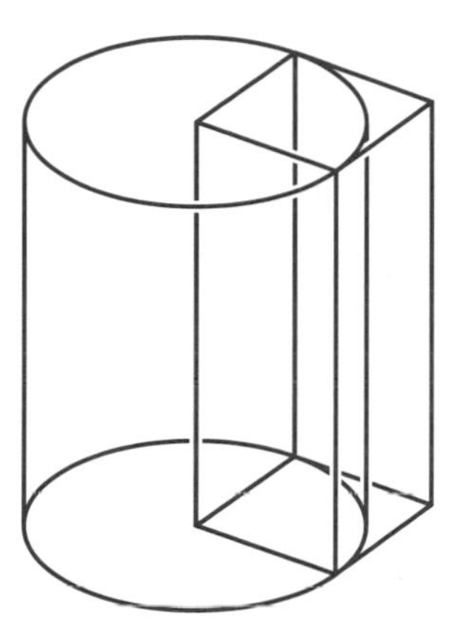

FIGURE 15. A set of cups containing a circular cylinder matched to four rectangular boxes of the same height whose bases form a square that encloses the circular base can be used to show that the volume of the cylinder is just a little bit more than the volume of three of the boxes. Hence the area of the circular base is just a bit more than three-quarters of the area of the corresponding square.

The idea of perimeter can be introduced by using a string or a belt, unmarked at first. The distance around a square tile is four times the length of the side of the tile, regardless of the size of the tile. If one circular disc has a radius twice that of another, then a string around the larger will fit twice around the smaller. A string around a disc will go around a square with sides equal to the radius a little more than three times. The crucial fact that the ratio of the circumference of the disc to the perimeter of the square is the same as the ratio of the volume of the cylinder to the volume of the surrounding box would be established only much later. But the fundamental idea that there is a fixed ratio between the perimeter of the disc and the perimeter of a square is something that every child should appreciate, long before any mention of the mysterious number π.

The relation between the area and circumference of a circle can be easily seen by cutting a circle like a pie and reassembling the pieces into a nearly rectangular shape. The area of a disc turns out to be equal to the area of a rectangle-like region with one side equal to the radius and the other equal to half the circumference (Figure 16). Subdividing the disc into more slices would make the correspondence even more exact. (Much later students will appreciate the limit concept hidden in this demonstration.) Unfortunately, there seems to be no such nice correspondence between the volume of a sphere and the volume of a rectangular box.

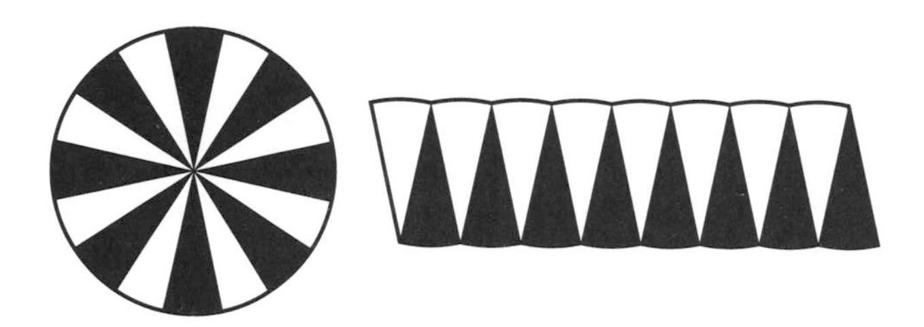

FIGURE 16. By slicing a circle into thin pie-shaped pieces and reassembling them into a rectangular-shaped region, children can readily see that the area of a circle is the radius (the height of the reassembled rectangle) times half of the circumference (the width of the rectangle).

VISUALIZING DIMENSIONS

Children in Froebel's kindergarten played with cubes and with subdivided cubes, squares and subdivided squares, and rods and subdivided rods (Figure 17). Eight small cubes fit together to form a large cube, twice as long, twice as wide, and twice as high. Four square tiles fit together to form a large square, twice as long and twice as wide. Two thin rods form a rod twice as long as the original.

Children at all levels can explore similar exercises. Here is a small cardboard box filled with sand, wrapped in paper, and tied with string. Here is another box—twice as long, twice as wide, and twice as high. How much more string do we need to tie it, or paper to cover it, or sand to fill it? It isn't necessary to have the ability to measure length or area or volume in order to experiment and find the answers: twice as much string, four times as many sheets of paper, eight times as much sand.

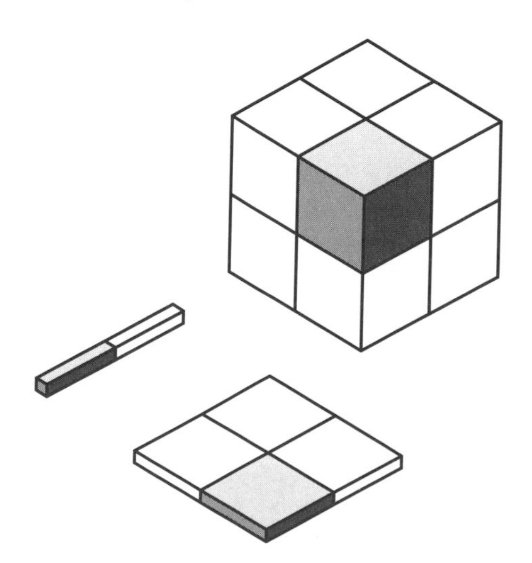

FIGURE 17. Nested cubes, squares, and rods illustrate the fundamental property of doubling factors: they represent the power of 2, depending on dimension.

These perceptions about changes of scale can take place even before the child has much experience with multiplication, and they can reinforce understanding of arithmetic processes.

Growth Factors

Children who first encounter changes of scale in the lower grades will recognize much later, when they learn about exponential notation, that doubling the size in dimension three leads to an increase in the volume of a factor of 2^3, whereas doubling the size of a two-dimensional square increases its area by 2^2. Whatever it might mean to have a box in four dimensions, exponents make very clear a pattern of doubling that predicts its size will increase by 2^4.

Each dimension, therefore, corresponds to its own growth exponent. A surprising fact is that there are geometric objects whose growth exponents are not whole numbers. These strange objects, which have a kind of "fractional dimension," are examples of a fascinating collection of geometric patterns known as "fractals." Since the creation of fractals usually requires a process that is applied an infinite number of times, it is only with the advent of modern computer graphics that it has been possible to carry out the experiments necessary to explore them effectively.

One of the earliest examples of a fractal was invented long before computers by the Polish mathematician Wacław Sierpiński. The first step in creating Sierpiński's figure is to remove a small triangle from the middle of a large one. The second step is the same as the first: remove the middle of each of the remaining triangles. Repeat this over and over again to obtain what is known as the "Sierpiński gasket" (Figure 18).

What's remarkable about Sierpiński's gasket is that doubling its size produces a figure that is composed of *three* copies of the original figure. This is very strange, because our experiments with tiles and cubes show that doubling factors are always powers of 2: if we double the size of something of dimension one, we get two copies of the original, whereas if we double the size of something of dimension two, we get four copies of the original. The Sierpiński gasket, therefore, must have a dimension somewhere between one and two—hence a fractional dimension. (Specifically, its dimension is the number d with the property that $2^d = 3$; this number d is the logarithm of three to the base two, namely 1.5849. . . .)

Fractals can be used to motivate a large number of mathematical discussions. Since they arise as a result of an infinite process, they can be discussed in relation to geometric series or repeating decimals. The unusual doubling properties of fractals give a geometric interpretation

FIGURE 18. This infinitely punctured triangle, known as Sierpiński's gasket, comprises three half-size copies of itself—not two or four as one would expect if its dimension were one or two. Hence it has a fractional dimension in between one and two.

for the logarithm to base two. Other fractal processes lead to figures like the Mandelbrot set, including some of the most striking examples of mathematical art.[5]

Rates and Averages

One of the most important skills we can give our students is the ability to interpret data geometrically. The geometry of area and volume can help students understand concepts like rates, accumulations, and average value. Here are three simple examples that illustrate this point:

- A driver travels at 40 miles per hour for 1 hour, then at 46 miles per hour for 2 hours. How far does she travel, and what was her average speed?

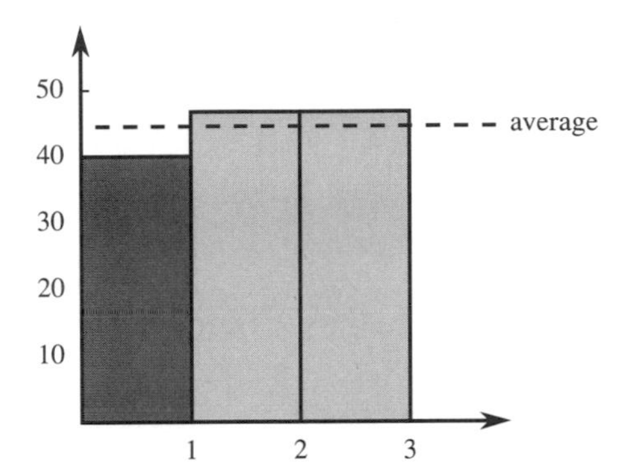

FIGURE 19. A bar graph geometrizes data from three similar problems and shows visually how the average corresponds to the height of a single rectangle with the same base and the same total area.

- A designer makes $40,000 a year for 1 year and then $46,000 for the next 2 years. What were his total earnings for that period, and what was his average salary?
- A fish tank is filled to a depth of 40 centimeters and two identical tanks are filled to a depth of 46 centimeters. What is the average depth of the water in the tanks?

All of these problems involve the same calculation, and all can be illustrated on the same diagram (Figure 19). In each case the total accumulation can be interpreted geometrically as the area of three rectangles. The average will be the height of a single rectangle with the same base and the same total area. It is also possible to graph the accumulation in a way that indicates exactly how many miles had been covered (or how much money had been earned) by a given time (Figure 20).

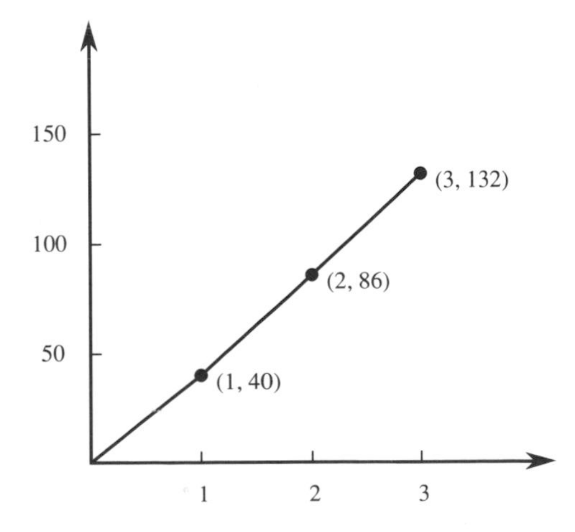

FIGURE 20. A linear graph displays the accumulation from three problems, indicating total miles covered or dollars earned. The relation between the corresponding bar and linear graphs is a precursor to calculus.

Each bar graph representation of rates (which mathematicians call a step function) leads to an accumulation graph formed from straight lines (i.e., a polygonal function). The process of finding the rate from the accumulation leads ultimately to the differential calculus, and finding accumulations from rates leads to the integral calculus. Although it is certainly not necessary for students to realize this connection as they develop their understanding of speeds and distances or salaries and earnings, every student can benefit from this type of mathematical experience both as preparation for calculus and as preparation for life.

Drawing Cubes

All children in Froebel's kindergarten practiced drawing. They played with drawing on one level while they learned on another. They learned to observe spheres, cylinders, and cubes; ultimately they learned to draw what they saw. In our day there is not as much emphasis on drawing, so we miss opportunities to develop the ability of our students to visualize geometrical relationships.

The most common way of representing a cube in most books is to draw a square, then translate it along an oblique axis (usually at a 45° inclination) and then to connect corresponding points (Figure 21). Although this is a perfectly valid representation of the structure of a transparent cube, no view of a cube actually looks like this image. Whenever we look at a cube, if one face appears as a square, then we must be looking directly toward that face; in this case the opposite face will be directly behind the face we see and not off to the side as it is in the traditional drawing. This is true whether we use a straight-down "orthographic" projection or foreshortening (Figure 22), where the back face appears smaller than the front.

Another popular method of drawing uses "isometric projection," which expresses three edges of a cube as segments of equal length

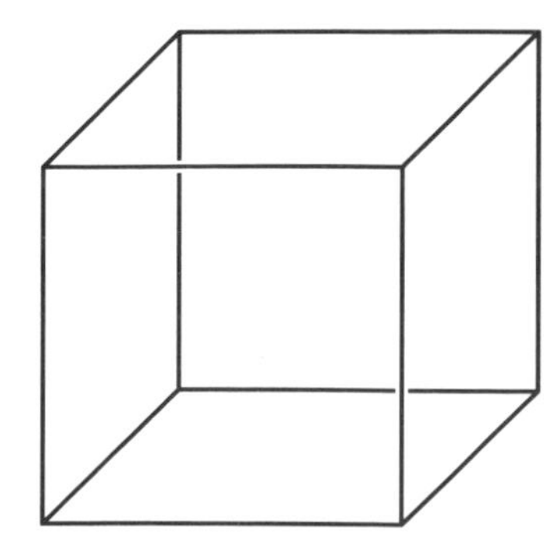

FIGURE 21. The typical representation of a cube as two identical squares with edges connected is quite unreal since no cube can ever appear just this way.

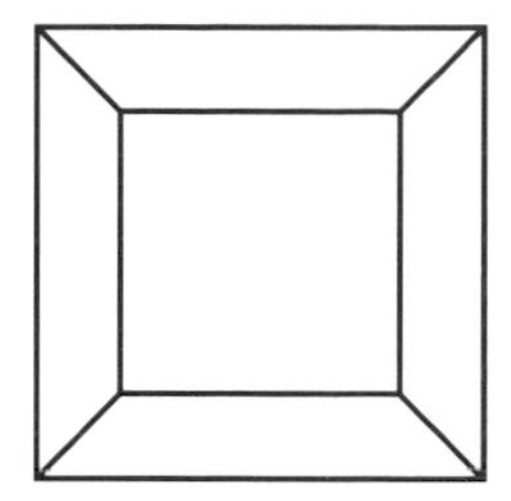

FIGURE 22. Two correct views of a cube are given by the "orthographic" projection (looking straight down) on the left or a foreshortened projection (on the right).

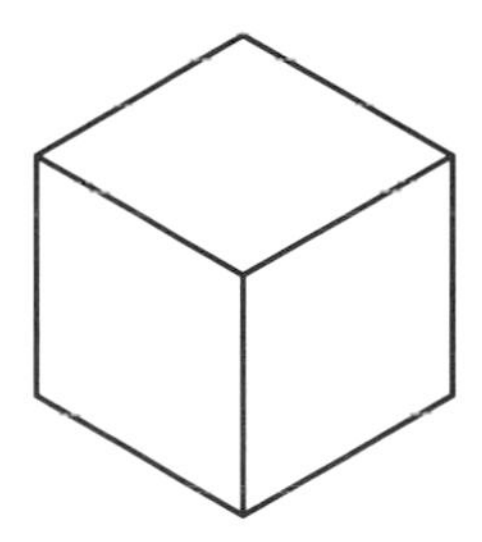

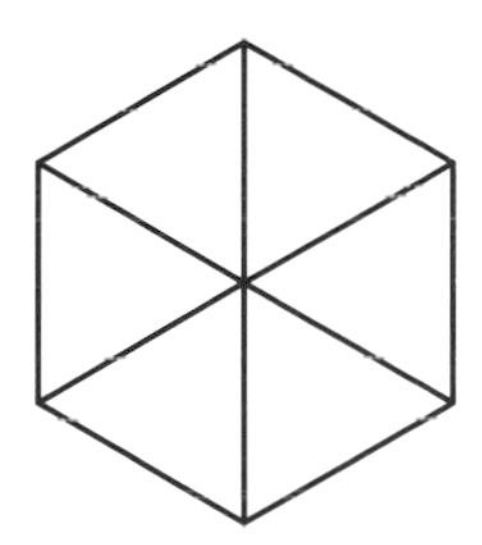

FIGURE 23. The symmetric "isometric" view of a cube, both solid and transparent, arise by looking at one corner at a 45° angle. Then the opposite corner lies directly behind the front corner, so only seven vertices are distinguished in this view.

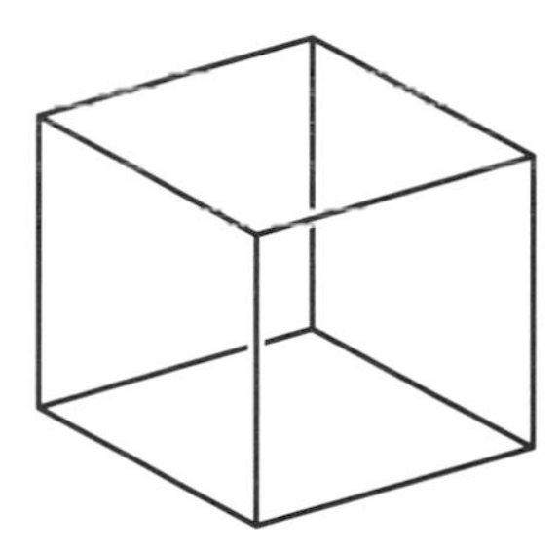

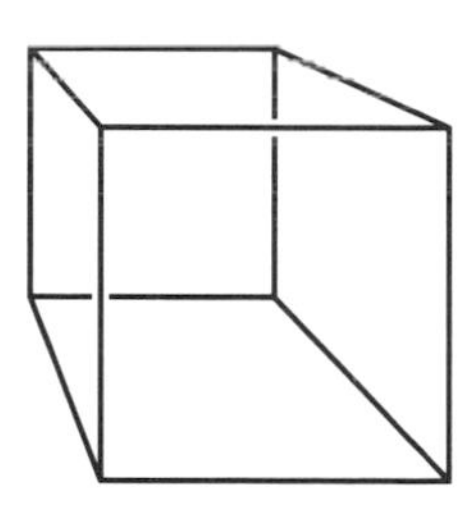

FIGURE 24. Two views of a cube in general position: orthographic (on the left) and one-point perspective (on the right).

meeting at 120° angles (Figure 23). This method has the disadvantage that two vertices of the cube are represented by the same point.

If we wish a more general image of a cube, we must draw each face as a non-square parallelogram (using a straight-on projection) or as a trapezoid (if we use one-point perspective) (Figure 24). The straight-on (or orthographic) projection is particularly easy to draw since the picture of a cube is completely determined once the position of the edges at one

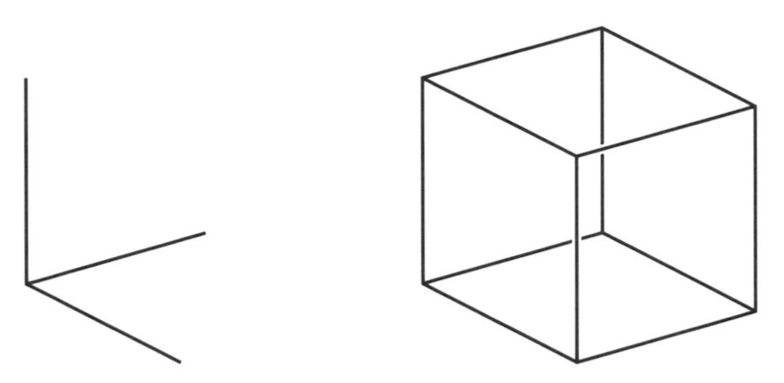

FIGURE 25. In an orthographic drawing, parallel lines in the cube are rendered as parallel lines on the page. Here the full orthographic view of a cube is determined by the orientation of the three edges at any corner.

corner is specified. In an orthographic projection, parallel edges of the cube appear as parallel edges in the image, so we can easily complete the picture once we know the position of the three edges at any corner (Figure 25).

Once we know how to represent a three-dimensional object on a two-dimensional page or computer graphics screen, we can go on to a much more complicated exercise, that of drawing a four-dimensional analogue of a cube, called a *hypercube* or *tesseract.* Many students encounter the idea of a four-dimensional cube in science fiction or fantasy literature, such as Robert Heinlein's story *... and He Built a Crooked House*[10] or Madeleine L'Engle's *A Wrinkle in Time*[14] or Edwin Abbott Abbott's *Flatland.*[1]

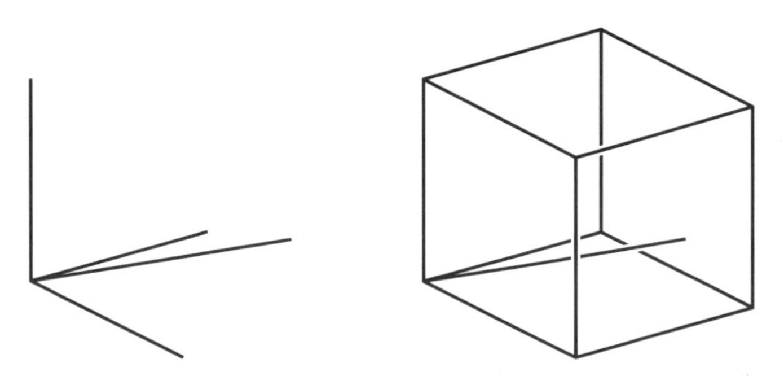

FIGURE 26. By adding a fourth direction to the traditional three-line corner that represents three-dimensional space, we lay a foundation for drawing four-dimensional objects. It shows the direction in which to move a cube to form a four-dimensional hypercube.

Usually a hypercube is constructed by moving an ordinary cube in a direction perpendicular to our space. Although we cannot actually achieve such a motion, we can still draw a picture of what such a construction would look like when the image is projected to a plane (Figure 26). We first draw the cube determined by three of the edges, then move a copy of the cube along the fourth direction and connect corresponding points.

The same procedure enables us to design a three-dimensional model of a four-dimensional cube, using sticks attached by clay balls (as suggested in the last century by Froebel) or more modern materials like drinking straws threaded together with yarn, or some standard building sets. Once again, the full image of a straight-on projection is determined as soon as we specify the four edges coming out of a point (Figure 27).

Just as a foreshortened view of a cube looks like a square within a square with corresponding corners connected, so the analogous foreshortened view of a hypercube looks like a "cube within a cube" with corresponding corners connected (Figure 28).

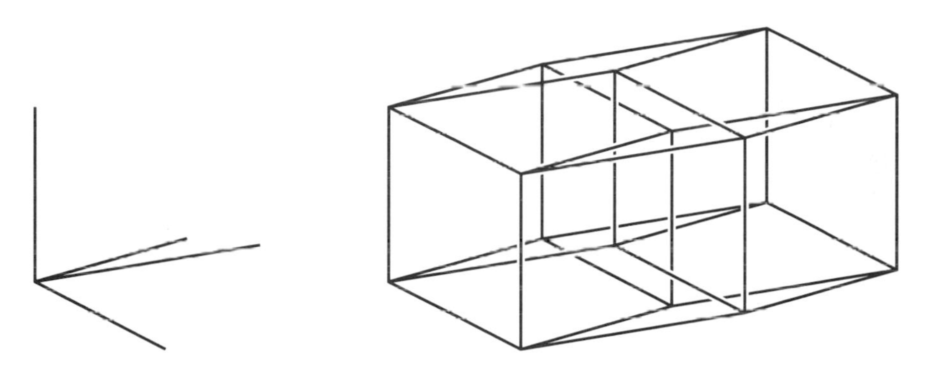

FIGURE 27. The completed hypercube formed by connecting corresponding vertices on two copies of a cube.

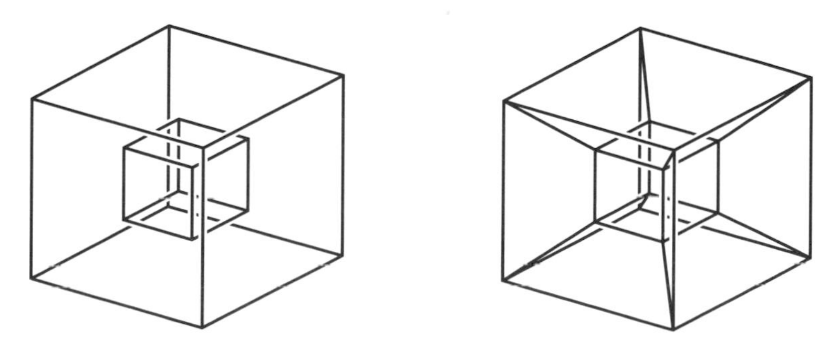

FIGURE 28. A foreshortened view of a hypercube, imagined as a cube within a cube with corresponding corners connected.

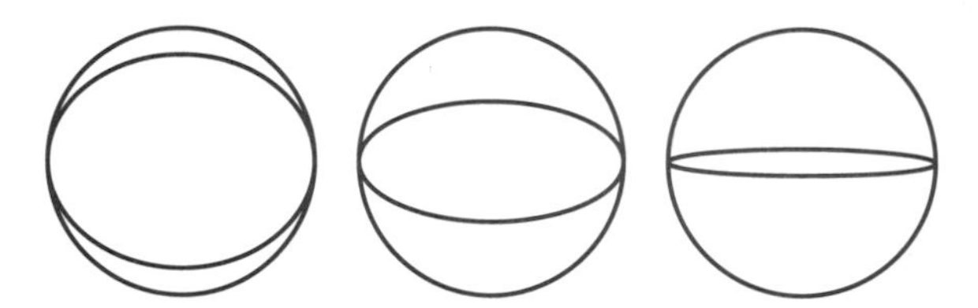

FIGURE 29. By rotating a sphere on which an equator has been drawn, it is easy to see that the images of a circle are always some type of ellipse.

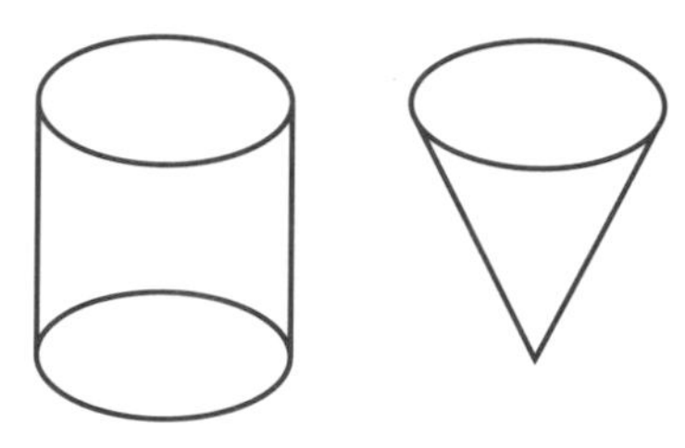

FIGURE 30. To draw cylinders and cones, one begins with an ellipse that represents a suitable perspective as the circular base.

A cube looks different from different perspectives. A sphere on the other hand always looks like a disc. Any way we look at it, it looks the same. If we mark an equator, then various views give images that are ellipses in different positions (Figure 29). Students also need to be aware of the basic principles of drawing these fundamental forms. It is a fact that a circle always looks like an ellipse, including the extreme case where the ellipse is still a circle or where it degenerates into a doubly covered straight-line segment. Observing this fact makes it easier to draw convincing cylinders and cones (Figure 30).

Modern computers are fast enough to produce a sequence of images showing different views of a rotating cube or hypercube, giving the illusion of a three-dimensional object. This process is very familiar to today's students who have grown up with computer-animated special effects and television commercials. We can make use of this experience to give students new appreciation for mathematical forms. As interactive programs become more widely available, students of all ages can have unprecedented opportunities, never before possible, to manipulate and explore geometric forms in three and higher dimensions.

COORDINATES IN DIFFERENT DIMENSIONS

One of the most important insights we can transmit to students at all levels is the utility of coordinate descriptions both to specify locations and to give instructions. Examples of coordinates can be made available at every stage of a child's development. There is no best way to develop understanding of different coordinate dimensions. You don't have to learn the first dimension completely before going into the second and then the third (and beyond). The invitation to examine coordinates from a dimensional standpoint is available at all times: we only have to make students aware of what they are seeing. Although experiences of different dimensions are always present, it is useful for our present analysis to separate phenomena according to the number of coordinates needed to locate a position or give an instruction.

Number Lines and Circles

Even at a very early age children can understand the significance of addresses. Anyone can appreciate the ordinary algorithm used for finding a specific location in terms of its street address: first go to the street, then find the number of some building. If it happens to be the one you are looking for, you are done. If not, go to a nearby building and check its number. If it is closer to the one you want, keep going in that direction. If it is farther away, go in the other direction. Stop when you get to the number you want.

Discussion of even this simple algorithm illustrates a number of important topics. We identify a location by a specific number, and we move along a one-dimensional path, in one direction or another, to get from one position to another. After the basic procedure is understood one can add refinements such as whether the address is on the even or odd side of the street. Estimating the distance one has to travel in order to get from one location to another is another refinement, leading to the geometric interpretation of subtraction as well as to the notion of absolute value.

Early exercises can take place on a number line with positive addresses or on real streets in a scavenger number hunt. Later the same notions can be used for scales with negative values, like temperature, where the vertical orientation of the thermometer emphasizes the directionality. "What happens when the temperature goes from 65 degrees to 40 degrees?" "It goes down by 25 degrees." Such observations can take place far in advance of introducing signed numbers.

Many cities use directional addresses in their street plan (e.g., in New York City there is both a West and an East 42nd Street). In this case the algorithm to find a building from knowledge of its address is slightly different but still easy enough to discuss at an elementary school level. The distance between two addresses on the same side is determined as usual, while the distance between two locations on different sides is the sum of their addresses. No memorization is required for such a statement! Signed numbers do not have to be mysterious.

A similar one-dimensional algorithm works in setting a clock, whether it is analogue or digital, depending on whether or not one can go backward as well as forward. Setting a watch is slightly different from finding a street address, even on a curving road that does not come back on itself. On a circular drive, however, the problem of locating a specific address is analogous to the problem of setting a watch: you can go in either direction and ultimately arrive at your destination. Of course one direction might be much easier than the other.

The problem of deciding on a strategy for locating an address on a circular drive is a good example of the kind of multistep problems that students should learn to attack. In this example, as in many others, there is no single answer—there are several strategies that will achieve the same result. The person facing the situation must decide first what the choices are and then what might be the advantages of each. The aim of minimizing effort is very easy to understand, easier than minimizing cost measured in money or some other quantity.

One-dimensional examples require just one number to locate any point. Directions for moving from one position to another are also one-dimensional: "Go three houses to the right" or "Go around counterclockwise five spaces" or "Go halfway around the circle to the opposite point." This last sort of instruction depends on the size of the circle and can form the beginning of an appreciation of angular measure.

Setting a clock, whether analogue or digital, provides an excellent example of "wrap-around." This phenomenon can also be viewed on a linear scale, for example, on the selector of many car radios. In many analogue devices the moving indicator stops at the extreme left or extreme right, while in the digital versions the indicator simply goes from the top value to the bottom. Finding a particular radio station then presents two different sorts of problems depending on the nature of the radio selector.

The dimensionality of gauges is an important concept that arises over and over again in mathematics as well as science. As students become more sophisticated in the kinds of numbers they use, they can introduce fractions or decimals into number lines and number circles. Locating a telephone pole along a road in a rural area requires a different kind of address, using fractions or real numbers representing actual distances. The numbers become more complicated, but the procedures remain the same.

Locating objects or addresses in a one-dimensional world can be accomplished efficiently by the bisection algorithm (or the variation of it that divides each interval decimally), a procedure with almost universal significance that is related, for example, to the informal technique used to find phone numbers. First you make a guess to divide your problem into two parts—by opening the phone book or by picking a number. Then you compare your guess with what you want and make a new guess that is in whichever part (above or below your first guess) that contains what it is that you are looking for.

A similar scheme can be used to find the "address" of the length of the diagonal of a square without requiring a calculator with a square root key. Finding the decimal equivalent of a fraction can be viewed as a more sophisticated version of the one-dimensional address problem. If we want to find 3/17, we can multiply different decimals by 17 to see if the product is bigger or smaller than 3. All decimals get put into one category or the other: it never comes out even. For 3/16 on the other hand one decimal does come out even, so there is a fixed location on the decimal line for the solution to this problem.

Lengths and Perimeters

The fundamental geometry problem for one-dimensional phenomena is the determination of distance along a path. Key examples include calculation or comparison of perimeters of curves and polygons. There is one geometric number—π—that all students should learn to understand.

Despite its universal significance, most people do not know how to answer when you ask what π is. Most lay persons respond with a numerical estimate, 3.1416 or 22/7, without knowing in either case whether this approximation is too large or too small. Mathematicians will give a definition in terms of a geometric property, usually something like "the ratio of the circumference of a circle to its diameter" or "the ratio of the area of a disc to the area of a square with side equal to the radius." The fact that these two ratios are the same is, of course, a major theorem of mathematics. One can get a tremendous amount of mileage out of a continuing discussion of the estimation of π, from the first time a kindergarten student realizes that the belt around a can reaches a little more than three times across the top, to second-semester calculus where one studies integrals for arc length.

Finding the circumference of a circle is a one-dimensional problem, so its answer should have a representative on the number line. But where is it? How can we determine whether or not a given number is less than this length or greater? Comparisons with the circumference of circumscribed and inscribed polygons is an effective strategy for dealing with these questions. Although such comparisons cannot determine π exactly, they can convincingly show whether 22/7 is slightly above or slightly below π.

Certain counting games are especially important for developing in children facility in the arithmetic of algebraic quantities. Students can choose instruction cards saying "move forward two spaces" or "back three" (F2 or B3), and they can follow the instructions with counters. Then they can be asked to trade two cards for a single card that accomplishes the same effect. By considering double or triple jumps, they gain experience with the idea of multiplying a signed number by a positive integer. The variations in the game are manifold. The operation of taking up three B4 cards from one's hand is the same as taking up one B12; putting down three B4 cards is the same as taking up one F12. One might introduce a symbolism: P3B4 = "Put down three B4 cards," which is the same as T3F4 = "Take up three F4 cards." Similarly, PB5 = TF5 and PF2 = TB2, yielding a complete algebra of transactions.

The pedagogical trouble with signed numbers is that we use them both for locations and for operations. The rule that "the product of two negative numbers is positive" is one of the earliest stumbling blocks that convince many students that mathematics means memorizing, not reasoning. Appropriate experience with counting games can restore intuition to the rules of negative numbers. Board games help students appreciate the value of scoring, first with simple addition (especially where movement depends on the throw of a pair of dice) and later in more complicated games where the score can be positive or negative. Scoring experiences are generally one-dimensional.

Planes and Surfaces

Children should become skilled in both following and giving directions. Any child should learn how to direct a person from one part of the school to another and perhaps to describe the neighborhood of the school. Although the algorithm for getting from one street address to another in an actual town might be quite complicated, an ideal town has a simpler structure. We can imagine a sequence of imaginary towns with different dimensional properties—a frontier town all stretched out along a single street or a village laid out on a rectangular plot. A model village could stimulate a good deal of the discussion, while a grid on which children could design their own town would allow for more variation.

No matter what the streets are named, we can still give directions on a grid by saying: "Go right two blocks, then turn left and go three blocks." For persons with a clear orientation, the instructions can be varied: "Go east two blocks, then north three blocks." The first instruction depends on the direction that the person is facing, and the second does not.

If the map of a village is hanging on a wall, we can use the natural coordinate directions: "Go right two blocks and up three." Certain pairs of instructions can then be combined: "Go left two, then up three" and "Go left three and down five" combine to give "Go left five and down two." By playing this game with cards, we can easily introduce the operation of adding ordered pairs and even of multiplying numbered pairs by positive integers. If we introduce "put" and "take" operations, we can extend the one-dimensional algebra of signed numbers to an algebra of two-dimensional quantities.

Notice that this algebra of instructions does not require the use of coordinates in the plane. The exercise carries additional value when addresses are given in terms of street numbers or compass directions. For one thing, this avoids the complications caused by negative numbers. To go from E3N4 to E7N2 requires a move of E4S2. The correspondence between this commonsense approach and the algebraic statement $(7,2) - (3,4) = (4,-2)$ is something that can come much later in a student's mathematical development. There are a great many people who are confused by negative numbers. They shouldn't bc.

"Taxicab geometry" provides an effective variation on the use of directional instructions. Students play the role of dispatchers, telling cabbies how to get from one location to another. "Just go three streets north and two avenues west" would be such a direction. The efficiency of the instructions—and the profit of the cab company—depends on many factors such as one-way streets, accidents, and traffic jams. One can easily imagine a board game that would model realistic city traffic and get students used to the idea of a two-coordinate instruction set.

The surface of the earth is another familiar example of a two-dimensional object. Even though it exists in three-dimensional space, we need only two numbers, latitude and longitude, to specify any location. A dispatcher of ships can give instructions to go 10 miles due east and then 5 miles due north. On the surface of the earth—but not on a flat plane—the order of these operations makes a difference: going 5 miles due north and then 10 miles due east can put a ship at a different position! The extent of this difference is an intrinsic indicator of curvature.

In teaching geometry we should not ignore the interactive video game. Today's students take for granted the fact that we can manipulate images on a two-dimensional screen by pushing buttons, turning dials, or twisting joysticks. Programs like LOGO offer students experience in giving simple geometric instructions to move points and objects around on a screen. This gives mathematics teachers a chance to introduce any number of important concepts, including repeated operations to form regular or star polygons and recursive processes for drawing fractal objects or space-filling curves.

Many video games employ wraparound, which introduces interesting ideas in different two-dimensional geometries. Frequently when a point is guided off the left side of a computer screen, it appears at the same height on the right side. This is analogous to the phenomenon on the digital radio dial, which just as well might be thought of as operating on a circle. A segment with its endpoints identified can be treated as a circle. Analogously, if we think of the points on the left side of a computer screen as identified with the corresponding points on the right side, then we are dealing not with a flat rectangle but rather with a cylinder.

But even more can happen. It is often the case that when a point moves off the top of the screen, it reappears at the corresponding position on the bottom, so we get a cylinder with its top and bottom identified. This gives a figure like an inner tube, which mathematicians call a *torus.* The geometry of a torus is in some ways like that of the plane, but in other ways it is very different. In the plane any polygon that does not intersect itself divides the plane into two pieces. But if we take a closed polygon that goes around the top of the torus, it does *not* separate the torus into two pieces: its inside is the same as its outside. Related to this phenomenon is the fact that on a torus we can find two closed curves that cross at exactly one point (Figure 31), whereas if two closed curves in the plane cross (not just touch), they must intersect in

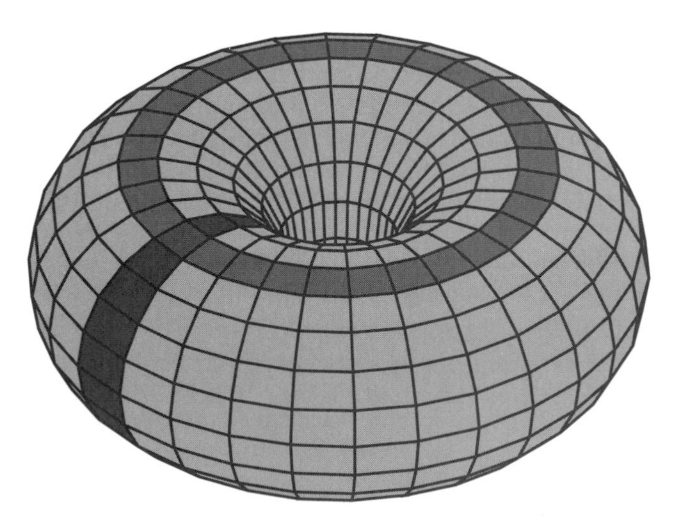

FIGURE 31. A torus, the mathematical name for a doughnut-shaped surface, is a two-dimensional surface in which two closed curves can intersect in just one point and in which a closed curve need not separate its inside from its outside.

an even number of points. An unusual object in many ways, the torus is ideal for keeping track of pairs of numbers from circles.

Three-Dimensional Space

It is a short step from two to three dimensions. From the two-dimensional village layout, we can move to the model of a city, where we have a height for each location as well as a position on the grid. We can augment taxicab geometry with elevator geometry. We specify a position by three numbers, for example, E3N4U9, referring to the ninth floor of a building at location E3N4. We can then determine an algorithm for getting from this location to E7N2U5. Note that in this particular geometry it makes a big difference in what directions one moves. The usual algorithm would be D9E4S2U5. Beginning with D4 gets you to the right level but in the wrong building! The situation would be different for a game played on a jungle gym, with instructions to move from one position to another by going a certain distance left or right, forward or back, up or down. In this case we can carry out the instructions in any order.

Another three-dimensional geometry arises if we want to specify the position of an airplane, giving its longitude, latitude, and altitude. Once again, it makes a difference in which order we give the numbers that indicate a given location or the directions for getting from one point to another.

Higher-Dimensional Spaces

The intuitions that students accumulate in dealing with coordinate pairs in the plane and coordinate triples in three-dimensional space lead naturally to coordinate geometry in higher dimensions. A thorough understanding of two and three dimensions provides an important foundation for the powerful generalizations of vector and matrix algebra in science and engineering, in economics and social science, and especially in computer science and graphics. We illustrate this progression with two examples.

The vertices of a square can be given by four points (0,0), (1,0), (1,1), and (0,1). To obtain the vertices of a cube, we can take the points of a square with zero in the third coordinate and then move the square one unit in the third direction to obtain four more vertices, with a 1 in the last coordinate:

$$(0,0,0),\ (1,0,0),\ (1,1,0),\ (0,1,0),$$

$$(0,0,1),\ (1,0,1),\ (1,1,1),\ (0,1,1).$$

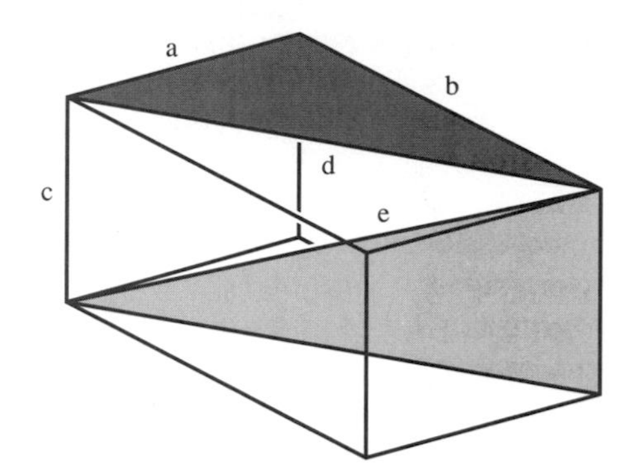

FIGURE 32. Generalizing the Pythagorean theorem to three dimensions by applying it to two different triangles found in a rectangular box.

Thus we can describe either the square or the cube as having vertices that are either 0 or 1 in each coordinate.

The procedure generalizes automatically: to obtain the vertices of a hypercube, we start with the eight vertices of a cube and put 0 in the final coordinate and then "move the cube in a fourth direction" to obtain eight more points with 1 the last coordinate:

$$(0,0,0,0),\ (1,0,0,0),\ (1,1,0,0),\ (0,1,0,0),$$

$$(0,0,1,0),\ (1,0,1,0),\ (1,1,1,0),\ (0,1,1,0),$$

$$(0,0,0,1),\ (1,0,0,1),\ (1,1,0,1),\ (0,1,0,1),$$

$$(0,0,1,1),\ (1,0,1,1),\ (1,1,1,1),\ (0,1,1,1).$$

We thus obtain the sixteen vertices of a hypercube, with 0 or 1 in each of four coordinates. It is this sort of representation that is ideal for communicating with a computer.

A second topic that generalizes in a very nice way is the Pythagorean theorem. If we think of this theorem as a way of calculating the length of the diagonal of a rectangle with given sides, then the extension to three dimensions is immediate: given a solid bounded by rectangular sides, we first apply the theorem to one side and then apply it to a rectangle built over the first diagonal (Figure 32). We easily get $e^2 = c^2 + d^2 = c^2 + (a^2 + b^2)$, so the length of the diagonal of a rectangular prism with sides a, b, and c is $\sqrt{a^2 + b^2 + c^2}$. The pattern is established, and the distance formula in four-dimensional space follows almost immediately. Students can then calculate the lengths of diagonals of the hypercube with the 0-1 coordinates. It turns out that the length of the major diagonal of a four-dimensional cube—say from (0,0,0,0) to (1,1,1,1)—is $\sqrt{4} = 2$, which is twice the length of a side.

CONFIGURATION SPACES

The coordinate descriptions that are so useful in giving locations and direction in familiar spaces of one, two, and three dimensions work equally well for phenomena whose specification requires more than three numbers. Exploratory data analysis, a statistical technique for dealing with these representations, is one of the most important applications of dimensions in current research. The ability to visualize and interpret multidimensional data sets may be one of the best gifts we can present our students in this modern age.

Some of the most useful and interesting examples of higher-dimensional phenomena occur as configuration spaces—collections of geometric objects representing certain structures or motions in the natural world. The most familiar spaces are the one-dimensional collection of points on a line, the two-dimensional collection of points in a plane, and the three-dimensional collection of points in space. But we can also consider the collection of lines in the plane, the collection of planes in space, the collection of all possible circles in a plane, or the collection of spheres in space. We illustrate this process by presenting several examples of phenomena that lead to higher-dimensional configuration spaces.

Consider the following (slightly unrealistic) situation: The lighting director of our local theater has to arrange a set of lights over the stage so as to illuminate certain parts of the floor at certain times. Sometimes the size of a spot is supposed to change during the course of a performance. Sometimes one colored circle is supposed to be contained in another. How can she keep track of all the circles of light and then design lighting directions so that an assistant can carry them out?

In this particular theater the lights all have the same form. A single bulb is suspended from a wire hanging down from the ceiling, and a conical shade directs the light out in a beam that meets the floor in a disc of light. The sides of the shade come down at a 45° angle, so the radius of the disc is equal to the height of the bulb above the floor (Figure 33). This makes it easy for the director to specify the location of any light, since she can indicate the position of the center of the disc using the same coordinates that the director of the play uses to give her instructions. That uses two coordinates, but the lighting director needs another number to represent the radius of the disc. She could, as an alternative, specify the height of the bulb above the floor, since in this idealized situation these two numbers are the same. Hence any particular disc can be represented by three coordinates, the first two

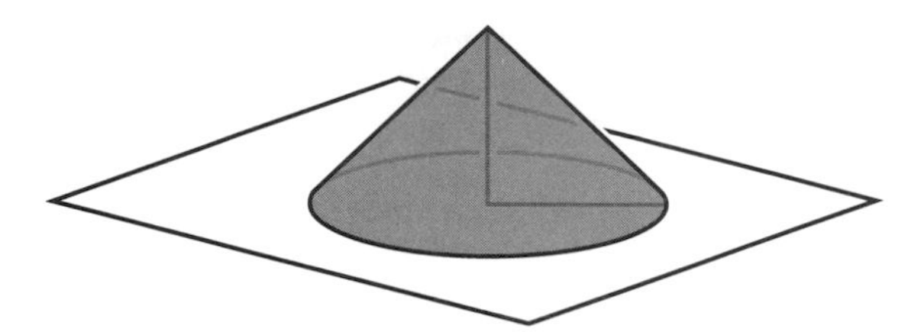

FIGURE 33. A spotlight with a shade set at a 45° angle will illuminate a spot on the floor of a stage whose radius equals the height of the light above the stage.

being the location of the center and the third giving the radius (or, in our special case, the height).

In this way we see that the collection of discs in the plane is three-dimensional; this collection is an example of a configuration space, each disc representing one element in the configuration of spotlights. To exploit the three-dimensionality as a bookkeeping device, the director can record the position of each light by giving three coordinates: for example, $(6, 8, 5)$ refers to the light with center at the $(6, 8)$ position on the floor and a radius (or height) of 5.

To call this a *space* indicates something more than convenience of recording. It is a signal that the arithmetic of the coordinates reflects properties of the geometry of lights. For example, a spotlight with coordinates $(6, 8, 5)$ stays on the stage, while the light $(6, 4, 5)$ shines off the front of the stage. It is easy to determine a rule to tell when a light stays away from the front rim of the stage, namely that the second coordinate be larger than the third.

More complex problems facing the lighting director can also be solved by referring to the coordinates. For example, when will one spot be entirely separate from another? In words, this happens when the distance between the points in the plane given by the first two coordinates is greater than the sum of the third coordinates. In symbols, the condition is expressed by $\sqrt{(x - x')^2 + (y - y')^2} > r + r'$.

In this configuration space the three coordinates do not play the same sorts of roles; so even though the geometry of the configuration space is three-dimensional, it treats the last coordinate differently from the first two. It is not identical to the usual geometry of ordinary three-space, where the Pythagorean theorem treats all coordinates the same way. An important aspect of configuration spaces are the special symmetries they possess.

The Fourth Dimension

Sooner or later everyone hears that time is the fourth dimension. That idea, however, limits the idea of dimensionality. Already in the last century writers realized that there are many situations in which time can be viewed as *a* fourth dimension, but by no means does it demand any special role as *the* fourth dimension. When physicists, especially relativity physicists, specify an event by giving three space coordinates and one time coordinate, they are using a four-dimensional configuration space. This space has its own geometry that is not the same as the geometry of four-dimensional Euclidean space, where distance is given by the generalized Pythagorean theorem. In the theory of relativity the distance between two events is given by the expression

$$\sqrt{(x-x')^2+(y-y')^2+(z-z')^2-(t-t')^2},$$

where time is measured in special units related to the speed of light.

The three-dimensional configuration space of spotlights provides a useful analogy for a four-dimensional space used in molecular modeling. The atoms that make up a molecule can be represented by small spheres of different radii. The description of a particular molecule, like the description of stage lighting, consists of a list of spheres of different sizes in different positions. Each sphere requires three coordinates to specify its center and one coordinate for the radius. Thus the configuration space of atoms is four-dimensional, and a molecule is a collection of such atoms arranged in a particular formation.

Using the language of the configuration space, we can describe a molecule to a computer and ask it to display different views. If we ask the computer to check that two atoms do not intersect, this involves an algebraic condition in four coordinates, namely

$$(x-x')^2+(y-y')^2+(z-z')^2-(r+r')^2>0.$$

The geometry of this configuration space is much closer to that of relativity theory than it is to ordinary Euclidean four-dimensional geometry. Interestingly it is this sort of question—avoiding intersections—that appears in the science of robotics, using large numbers of coordinates to keep track of objects moving through configuration spaces of high dimension.

Suppose each light on our sample stage possesses a rheostat that can control the current—hence the brightness—of the spot. If we add brightness to the coordinates of the spotlight, then the configuration space will be four-dimensional. If we want to encode the color of each spotlight as well, then the dimensionality jumps again. The specification of color requires three more coordinates representing either hue, saturation, and value or the relative amounts of red, yellow, and blue (for pigments) or red, green, and blue (for lights). So the lighting director will now have seven coordinates for each spotlight: two for floor position, one for radius, one for brightness, and three for color. Thus even a simple example can lead to a configuration space of high dimensionality.

Relativity physics began by considering four-dimensional collections, with three dimensions for space and one for time. Recently modern physics has become much more complicated. Some current models keep track of seven dimensions that act like space and four that act like time, to give an 11-dimensional configuration space. Another important model uses a configuration space with 26 dimensions. In each case the choice of the model depends to some degree on the kinds of mathematics that apply in these dimensions, as an aid to keeping track of the complex interrelationships among events in these high-dimensional spaces.

Statics and Dynamics

Here's yet another type of configuration space, set up by a simple story. For the school sculpture show two students want to decorate the back wall of the hall with a pattern of plastic strings. They decide to stretch them from the left-hand edge of the wall down to the floor. By trial and error the week before the show, they come up with a pleasing design, using more than twenty strings. They can't leave them up until the show so they have to find a way of recording the positions so they can put them up again later. How many numbers do they need to specify the position of each string? What is the dimensionality of the collection of strings?

It is easy to see that the dimensionality of this configuration space is two: it takes just two marks to locate a given string, one along the floor and one up the left edge of the wall, and each of these locations can be specified by a single number. The pair of numbers (4, 3), for example, could represent the string that goes from the point four feet over on the floor to the point three feet up on the wall edge (Figure 34). The collection of pairs, one pair for each string, tells the positions of all strings. It is even possible to record these ordered pairs in a specific sequence so the students will know which order to follow when they replace them.

In a way this coding is like the old game of "connect the dots" where a polygon is determined by a sequence of ordered pairs, so by connecting

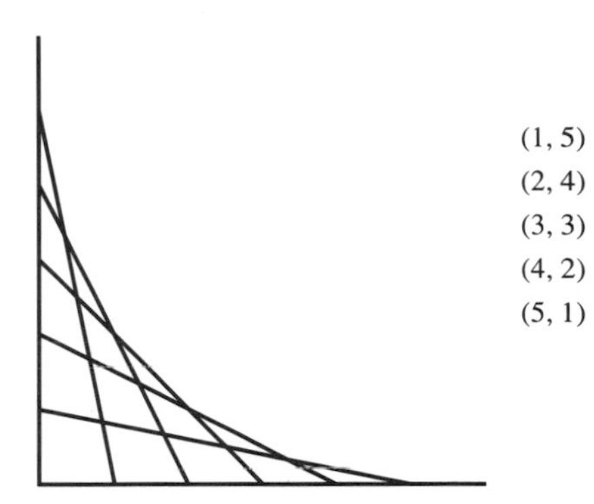

FIGURE 34. A configuration space of two dimensions can represent the positions of strings that run from the floor to locations on the left edge of the wall.

the dots in order we draw the polygon. In our sculpture story the basic elements are not points but segments: by forming the sequence of string segments, we re-create the wall sculpture.

If we increase the dimensionality of the configuration space, we can allow the bottom of the string to be placed anywhere on the floor, with the top still somewhere on the left edge of the wall. We still need one number for the height, but now the record will have to include two numbers for the floor coordinates. The collection of segments would then be three-dimensional, yielding greater possibilities of more interesting sculptures.

By allowing the strings to start anywhere on the vertical wall and end up anywhere on the floor, we would have a realization of a four-dimensional system. Simple algebra would then enable one to predict, for example, whether or not two strings are going to intersect. When we are laying strings along a wall, it is commonplace for them to intersect. Such intersections are rare if we are in a three-dimensional collection and rarer still for the four-dimensional system of segments in space. It is also interesting to look for configurations of segments that correspond to familiar configurations in ordinary space. What collection of segments in a two-dimensional configuration space corresponds to a line joining two points? What segments in a three-dimensional collection correspond to a coordinate plane in three-space? Questions such as these can yield striking and unpredictable visual effects in the string sculpture.

The dimensionality of a configuration space becomes especially important when we consider dynamic problems. When a point is moving on a line, we can describe its *state* at any given time by giving two numbers, one for its position and a second for its velocity. The state space is therefore two-dimensional, and a point moving according to a given physical law, like a ball bobbing up and down on a spring, will describe a curve in that state space. Similarly a point moving in a circle, like a swinging pendulum, will have a two-dimensional state space giving its angular position and angular velocity.

The state space of a point moving in a plane will be four-dimensional, with two points for location and another two for velocity. Scientists analyzing the motion of a satellite have to work in a six-dimensional state space, with three coordinates for position and three for velocity. The laws of physics will restrict the actual states of a system to some lower-dimensional space. Indeed, scientists devote a good deal of effort to analyzing the shapes of these spaces. For example, the motion of two pendulums corresponds to a curve on a torus in four-dimensional space. The study of such high-dimensional dynamical systems is an extremely important subject in modern applied mathematics.

SLICING IN DIFFERENT DIMENSIONS

When Froebel presented his geometric gifts, he did not want them to appear static. One of the first gifts was a display of three basic forms suspended by strings in various ways (Figure 35). As the objects rotated,

FIGURE 35. Froebel's kindergarten included basic shapes that could be hung from eyelets at different positions, then viewed from different perspectives to see various cross-sectional shapes.

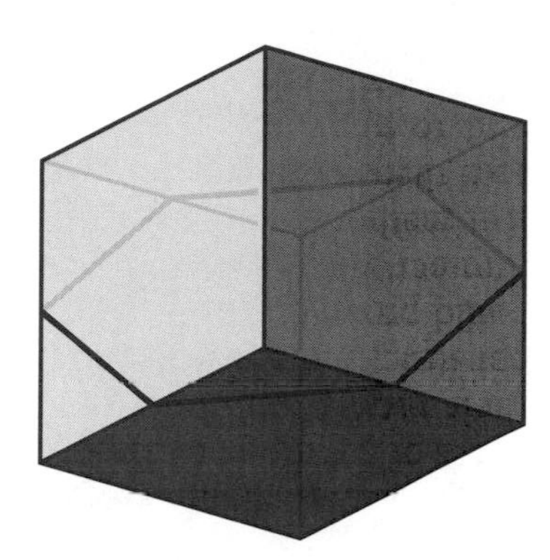

FIGURE 36. The central diagonal cross section of a cube turns out to be a regular hexagon whose six edges cut off triangles on each of the six faces of the cube.

children could observe them from different views and ultimately come to an appreciation of their symmetries and structures.

In the model devised by Froebel, the sphere, the cylinder, and the cube all had eyelets attached so that they could be suspended in different ways. Because of its symmetry, the sphere had only one eyelet. The cylinder had three: one in the center of an end disc, one in the center of a side, and one on the rim. The cube also had three: one in the center of a face, one in the center of an edge, and one at a vertex.

The various views of these rotating objects lead to one of the most intriguing exercises in understanding forms in space, namely the determination of cross-sectional slices. One way to visualize this without actually applying a knife to a real model is to imagine what would happen if we gradually submerged the block in water. How will the *shape* of the water level change?

The exercise that is most difficult for students is to visualize the shape of the "equator" of a cube suspended from a vertex. A student who has looked carefully at a real cube will have a much better chance of figuring out that the answer is a hexagon (Figure 36). This fact can be demonstrated nicely by stretching a rubber band around a cube. A cardboard model for the pieces of this decomposition of a cube can be made by cutting corners from three squares and placing them on the sides of a regular hexagon (Figure 37).

A transparent plastic cube half filled with a colored fluid can be manipulated to show the various slices through the center. If the cube is exactly half full, the shape of the liquid's surface will always be a central slice—that is, a slice through the center—regardless of the cube's orientation. It is a good challenge to then ask students to figure out which position of the cube produces the central slice with the greatest area. (It is not the hexagonal slice!)

Already in the last century when Milton Bradley took up the manufacture of Froebel's kindergarten materials in the United States, he

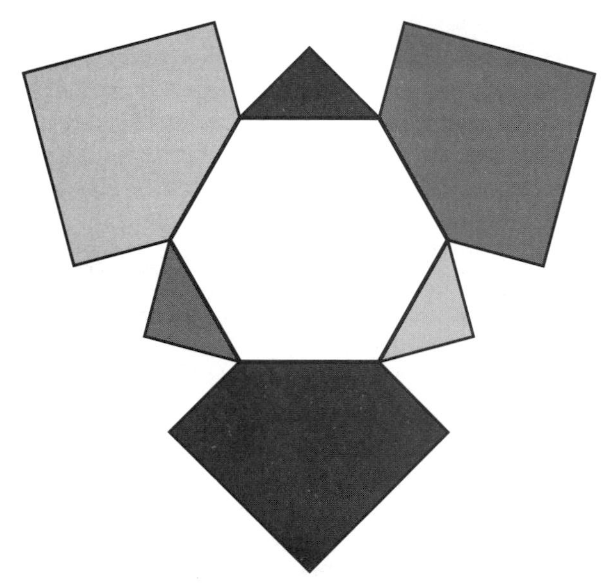

FIGURE 37. By folding this template into a solid figure, one gets half of a cube sliced on the central diagonal. Two such solids can be reassembled to form the cube by placing the hexagon faces together.

included in one of his sets another figure—a cone. The conic sections are phenomena that can be seen and appreciated long before students are introduced to analytic geometry. Once again, a transparent cone partially filled with liquid can illustrate the changing conic sections as the object rotates.

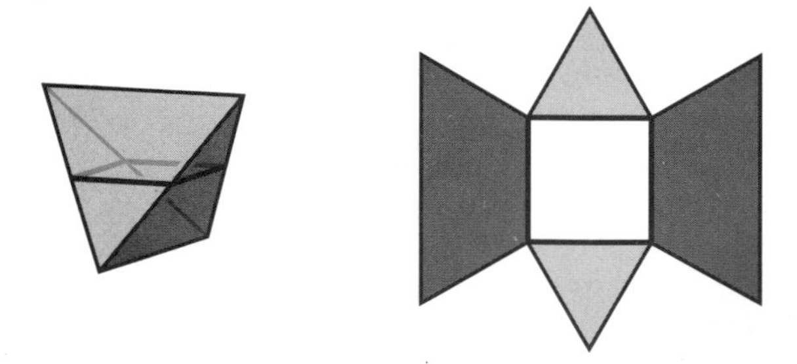

FIGURE 38. As the central slice of a six-sided cube yields a regular six-sided polygon, so the central slice of a four-sided tetrahedron yields a regular four-sided polygon—that is, a square. The template on the right provides the means for constructing half of a tetrahedron; two such pieces make an excellent geometric puzzle.

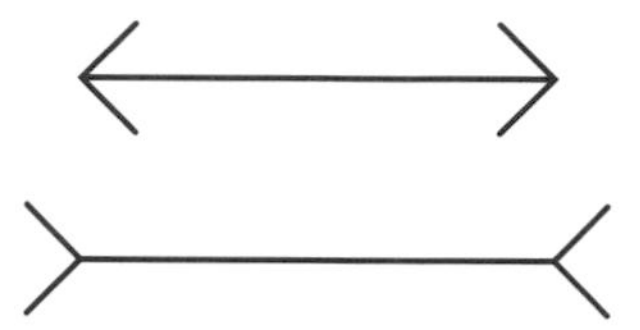

FIGURE 39. Appearances can be deceiving: the direction of the arrowheads changes the apparent length of the lines without changing their actual length.

The investigation of slices of polyhedral objects leads to an interesting puzzle. If we slice a triangular pyramid by a plane parallel to one of its faces, we get a series of triangles. If we slice by planes parallel to one of the edges, we get rectangles, and in the central position, a square (Figure 38). Students can make cardboard polyhedral models of the two pieces of this decomposition by cutting and folding an appropriate pattern. Many people find it very difficult to put these two identical pieces together to form a triangular pyramid. The difficulty seems to be a three-dimensional analogue of the optical illusion that makes two lines of equal length seem different if we put arrows on the ends (Figure 39).

Visitors from Higher Dimensions

Over one-hundred years ago Edwin Abbott Abbott used slicing to illustrate the dimensional analogy in his classic satire *Flatland.*[1] It is a great exercise to try to take on the viewpoint of A Square, living in a two-dimensional universe, especially when he is visited by a sphere from a higher dimension. The frustrated attempts of the sphere to teach A Square about the third dimension give wonderful insights into the challenges of communication and visualization in geometry. (Early parts of *Flatland* may be difficult for some students, and some of the social satire may be skipped over at first reading. Abbott was an active education reformer and worker for equality who was satirizing the narrow-minded attitudes of Victorian England with respect to class society and particularly with respect to women. Only at the end does A Square begin to gain a more enlightened view of his society.)

What would happen if we were visited by a sphere from a dimension higher than our own? Instead of growing and changing circles in a plane, we would see growing and changing spheres in space. We would be inclined to interpret such an event as the inflation and deflation of a balloon, but the point of the exercise is that such a phenomenon could be interpreted equally well as the slices of a hypersphere penetrating our three-dimensional universe.

If A Square were visited by a cube from the third dimension, he would see a variety of polygons, depending on the position of the cube as it passed different water levels. What would be the analogous three-dimensional slices of a four-dimensional hypercube? This is one place where computer graphics can be of great help (as in the film *The Hypercube: Projections and Slicing*).[3]

Slicing techniques are important in many modern scientific applications, especially since the development of computer graphics. X-ray tomography uses computer graphics in the reconstruction of three-dimensional objects from planar sections. Topographers and geologists construct and analyze contour maps showing the elevations of different configurations above and below the surface of the earth. Similar slicing methods are used by biologists, while researchers in materials science use computer graphics to show the parts of a three-dimensional surface with a given temperature or density. Exploratory data analysis uses techniques of projections and slicing to investigate high-dimensional data sets from social sciences as well as from the physical and biological sciences.

Students of calculus will appreciate the power of slicing techniques—for example, in relating the volume of a surface of revolution to the changing areas of its circular cross sections or in finding the contour lines on the surface of a graph in three-space. Long before students are introduced to the notions of critical point theory, they can already understand and appreciate slicing phenomena that relate different dimensions. What happens if we slice a doughnut or a bagel in different directions? It is easy to carry out the actual experiments and see that there are positions where the slice yields a pair of circles. Less obvious is the slice that consists of two interlocked circles. Again, a good way to see this would be to experiment with a transparent inner tube filled halfway with colored liquid. Geometry can be a surprising observational science.

COUNTING COMBINATIONS

Many combinatorial and algebraic questions arise in the investigation of geometric figures; these can be introduced at different educational levels, right up to the frontiers of research. How many edges does a triangular pyramid have? We can follow Froebel's suggestion and make a model out of toothpicks and peas, then count the edges. Or we can simply draw a picture of the object (Figure 40) and count the six edges.

The procedure for drawing such a diagram suggests an algorithm for determining the number of edges. Start with a point, then choose a distinct point and draw the one edge connecting it to the one we already had. Now choose a new point and connect it to the previous two points to get two more, for a total of three. (We have to be careful not to

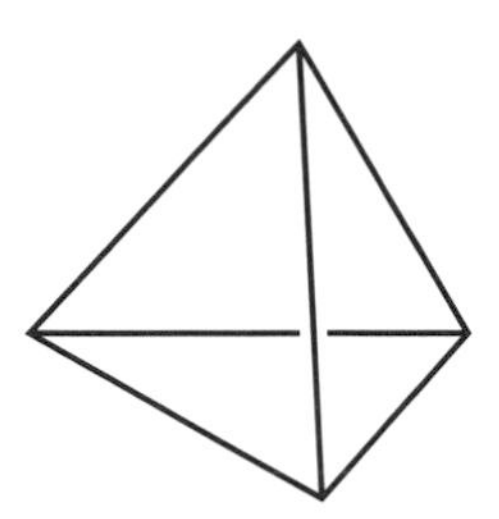

FIGURE 40. The tetrahedron—the simplest regular polyhedra—has four triangular faces, six edges, and four vertices.

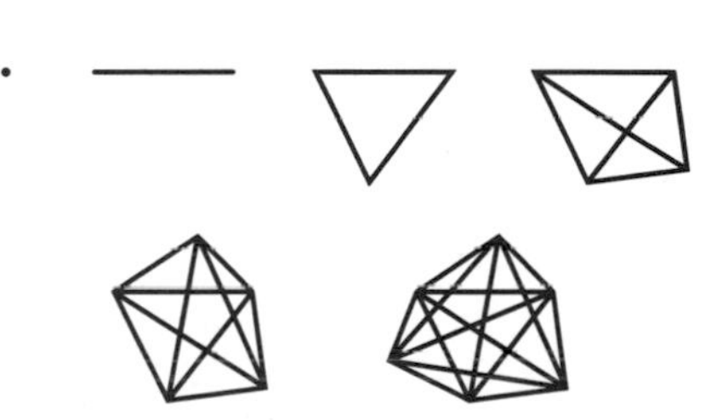

FIGURE 41. By adding one new point with line-segment connections to each previous vertex, one can construct in sequence the complete graphs on 1, 2, 3, 4, 5, 6, ... points.

choose the new point on the line containing a previous edge.) Next choose a new point not lying on any of the three lines determined by the edges already constructed, and then connect this new point to the previous three. This yields three new edges, for a total of six.

We can repeat this process to draw the figure—called a *complete graph*—determined by five points (Figure 41). First choose a point not on any of the six lines containing previously constructed edges, and then connect it to the previous four points to obtain four new edges—for a total of 10. A similar construction can produce the complete graph on six points and more if so desired.

What is the pattern that emerges from this procedure? It becomes apparent if we arrange the results in a table:

Number of points:	1	2	3	4	5	6
Number of edges:	0	1	3	6	10	15

In each case the number of edges is the number of pairs of points, which leads directly to the study of combinations. Based on the sequence of construction, it is easy to see that the number of edges at stage n is the sum of all numbers less than n. For example, the number of edges formed by six points is $1+2+3+4+5 = 15$. Some students may know the formula $n(n+1)/2$ for the sum of the first n integers, perhaps in conjunction with the famous story of the young Gauss who used this formula to add up all the numbers from 1 to 100. Another type of pattern is revealed by the table—that the number of edges at any stage is the total of the previous number of edges and the previous number of vertices.

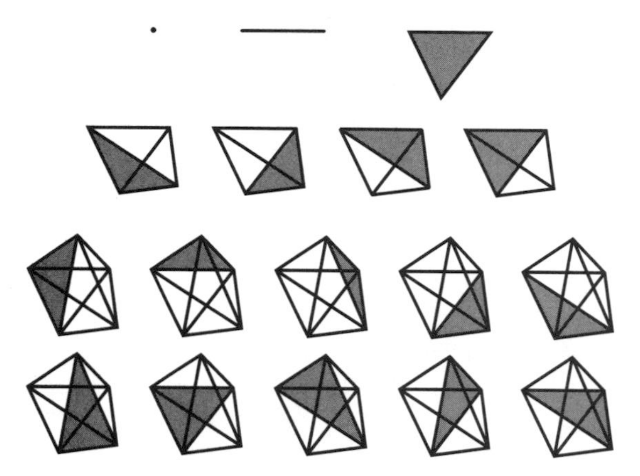

FIGURE 42. A display of different triangles determined by complete graphs shows that every subset of three vertices determines a triangle. Hence counting triangles is equivalent to counting triples of vertices.

Counting Triangles

Spatial perception tests often ask students to extract a simple figure from a complicated one. Counting edges is one of the simplest of such tasks. Next in difficulty would be counting the number of distinct triangles (Figure 42). By marking each triangle we can extend our table to include the new information:

Number of points:	1	2	3	4	5	6
Number of edges:	0	1	3	6	10	15
Number of triangles:	0	0	1	4	10	?

To fill in the missing value we can reason from patterns, many of which are just like those that relate edges to points. Since there are as many triangles as there are distinct triples of vertices, the total number of triangles is just the combinations of a certain number of objects taken three at a time. Alternatively, as before, we can use a recursion relationship: the number of triangles at any stage is the sum of the previous number of triangles and the previous number of edges. The latter is the easiest to calculate: it shows that the number of triangles that can be formed from 6 points is 20. [In general the number for n points is $n(n-1)(n-2)/6$.]

Students who have studied some algebra will be able to relate these numbers to the binomial coefficients:

$$(a+b) = a+b$$
$$(a+b)^2 = a^2 + 2ab + b^2$$
$$(a+b)^3 = a^3 + 3a^2b + 3ab^2 + b^3$$
$$(a+b)^4 = a^4 + 4a^3b + 6a^2b^2 + 4ab^3 + b^4$$
$$(a+b)^5 = a^5 + 5a^4b + 10a^3b^2 + 10a^2b^3 + 5ab^4 + b^5$$
$$(a+b)^6 = a^6 + 6a^5b + 15a^4b^2 + 20a^3b^3 + 15a^2b^4 + 6ab^5 + b^6$$

Removing the literal factors leaves a shifted version of Pascal's triangle:

1	1					
1	2	1				
1	3	3	1			
1	4	6	4	1		
1	5	10	10	5	1	
1	6	15	20	15	6	1

The fourth row, for example, gives in succession for $n = 0, 1, 2, 3$, and 4 the numbers of objects with n vertices formed from the four points: dots, lines, triangles in the middle, with the empty set and the whole set at the ends (where $n = 0$ and $n = 4$).

Observant students may see another important pattern—that the sum of any row is a power of 2. There is a sophisticated way of stating this observation: the sum of the numbers of simplices of different dimensions in an n-simplex—including the whole object and the empty simplex—is 2^{n+1}. This same relationship can be observed by setting both $a = 1$ and $b = 1$ in the table of binomial expansions or by relating the binomial coefficients to the combinations of $n + 1$ elements taken $k + 1$ at a time. The total number of possible combinations is then 2^{n+1}, the total number of subsets chosen from among $n + 1$ elements. This basic counting argument can motivate many topics in elementary probability.

Counting Squares and Cubes

Similar observations emerge if students investigate the numbers of vertices, edges, and faces of cubes and hypercubes in various dimensions. Just as there is a hierarchy of subsimplices within each simplex, there is an analogous sequence of squares and cubes within each n-dimensional cube. A 3-cube has 8 vertices, 12 edges, and 6 squares, as can be verified by an actual count. A square, or 2-cube, has 4 vertices, 4 edges, and 1 square. A 1-cube is a segment with 2 vertices and 1 edge,

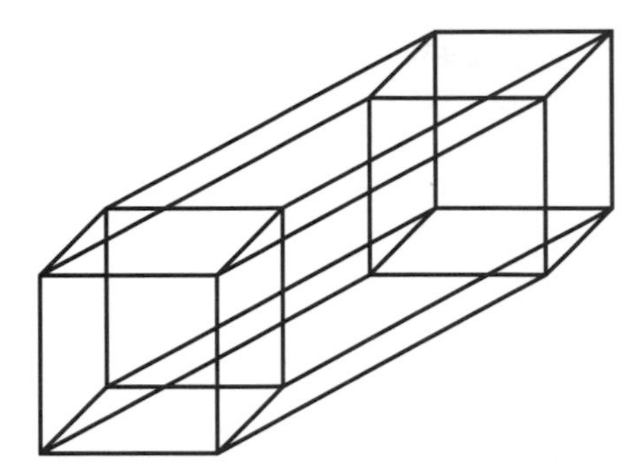

FIGURE 43. Framework for hypercube: two cubes with joined edges yield 16 vertices and 32 edges.

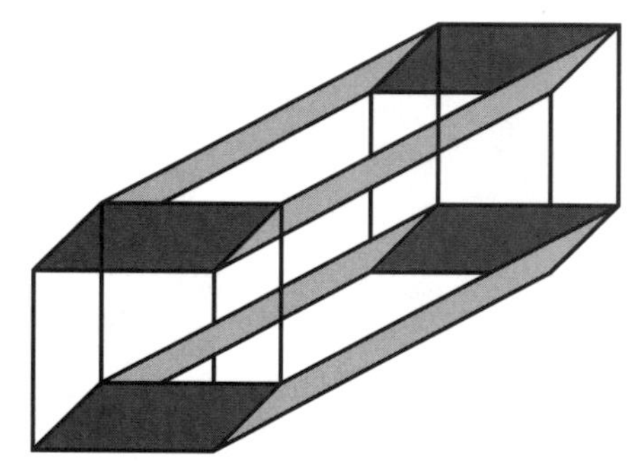

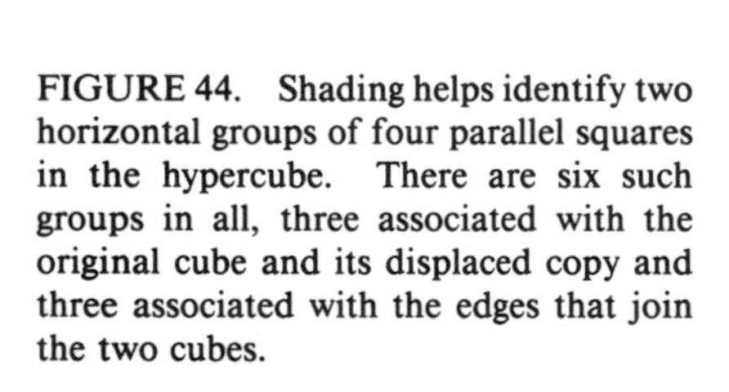

FIGURE 44. Shading helps identify two horizontal groups of four parallel squares in the hypercube. There are six such groups in all, three associated with the original cube and its displaced copy and three associated with the edges that join the two cubes.

and a 0-cube is point with 1 vertex. This data can form the beginning of another table:

DIMENSION:	0-cubes (points) Vertices	1-cubes (lines) Edges	2-cubes (squares) Faces	3-cubes (cubes) Cubes	4-cubes (hypercubes) 4-Cubes
Point:	1	0	0	0	0
Line:	2	1	0	0	0
Square:	4	4	1	0	0
Cube:	8	12	6	1	0
Hypercube:	16	?	?	?	1

When we try to fill in the missing numbers for a hypercube, the process becomes a bit more difficult. We know how to generate a hypercube—move an ordinary cube in a direction perpendicular to itself. As the cube moves, the 8 vertices trace out 8 parallel edges. This yields 12 edges on the original cube, 12 on the displaced cube, and 8 new edges traced by the movement for a total of 32 edges on the hypercube (Figure 43).

Counting squares presents more of a problem, but a version of the same method can be used to solve it. First observe that there are 6 squares on the original cube and 6 on the displaced one. To these 12 we must add the squares traced out by the edges of the moving cube.

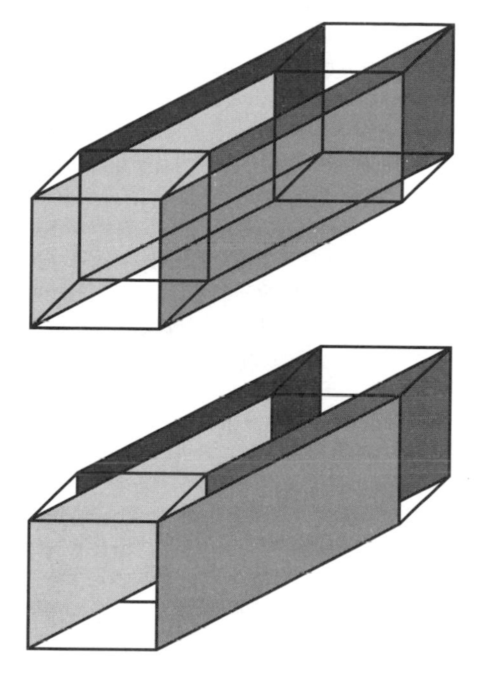

FIGURE 45. A group of four vertical squares in a hypercube determined by the horizontal displacement of the original cube. These squares are easier to see when background lines are removed, as in the lower figure.

It helps to group edges and squares in parallel bundles. The edges in the hypercube come in four groups of 8 parallel edges. Similarly the squares can be classified in four groups of 4 parallel squares, one such square through each vertex. Two horizontal groups are rather easy to see (Figure 44); another group of four vertical faces become clearer when we remove some of the extraneous lines (Figure 45).

Student teams can easily identify the remaining three groups of four squares. It is easier to do this when the four squares do not overlap and relatively more difficult when the overlap is large. The entire set consists of 24 squares.

Grouping edges or faces is particularly effective when an object possesses a great deal of symmetry, as does the hypercube. We can study the relation between symmetry and grouping by looking at different dimensions. Symmetries of a cube, a square, or a segment arise by permuting the edges at each vertex in different ways and by moving each vertex to another position. The collection of all symmetries of the cube or hypercube is an important example of a *group,* an algebraic structure that reflects geometric properties. The symmetry group of a cube is the collection of permutations of its vertices that preserve its structure. The attempt to codify the relation of permutations to symmetries of algebraic and geometric structures provided considerable impetus for the development of modern algebra during the past two centuries. Even now symmetry groups continue to fuel theoretical work in atomic physics.

The crucial observation about the hypercube is that it is so highly symmetric that every point looks like every other point: if we know what happens at one vertex, we know what happens at all vertices. For example, at each of the 16 vertices of the hypercube there are 4 edges, for a total of 64. But this process counts each edge twice, so the actual number of edges is half of 64, or 32.

At each vertex there are a certain number of square faces. How many? As many as there are ways to choose two edges from among the four edges that meet at the vertex. Once we have chosen one edge from among the four, there remain three possibilities for the second; together, these yield 12 pairs. As before, each pair of edges appears twice in this list, once in each order. So these 12 pairs yield 6 different squares at each vertex. All 16 vertices together then yield 96 squares. But each square is counted four times, once for each of its vertices. Hence the true total is $96/4 = 24$ squares in a hypercube. This reasoning confirms the direct count of six groups of four squares that we saw in drawings of the hypercube, but it is reached by a method that would work even if applied to a five-dimensional cube.

Seeking Patterns

Advanced students can express these results in a general formula. Let $\square(k, n)$ denote the number of k-cubes in an n-cube. To calculate $\square(k, n)$ we begin, as before, by counting how many k-cubes there are at each vertex. Each k-cube is determined by a subset of k distinct edges from among the n edges emanating from each vertex. Therefore the number of k-cubes at each vertex is $C(k, n) = \binom{n}{k} = n!/k!(n-k)!$, the combination of n things taken k at a time. Since there are $C(k, n)$ k-cubes at each of the 2^n vertices, the total number of k-cubes appears to be $2^n C(k, n)$. But in this count each k-cube is counted 2^k times, so we divide by that number to get the final formula: $\square(k, n) = 2^{n-k} C(k, n)$.

Remembering the pattern of powers of 2 that come from the sums of rows in the simplex table, we naturally seek a similar pattern for cubes. In this case the entries in each row add up to a power of 3:

DIMENSION:	0-cubes (points) Vertices	1-cubes (lines) Edges	2-cubes (squares) Faces	3-cubes (cubes) Cubes	4-cubes (hypercubes) 4-Cubes	Sum
Point:	1	0	0	0	0	1
Line:	2	1	0	0	0	3
Square:	4	4	1	0	0	9
Cube:	8	12	6	1	0	27
Hypercube:	16	32	24	8	1	81

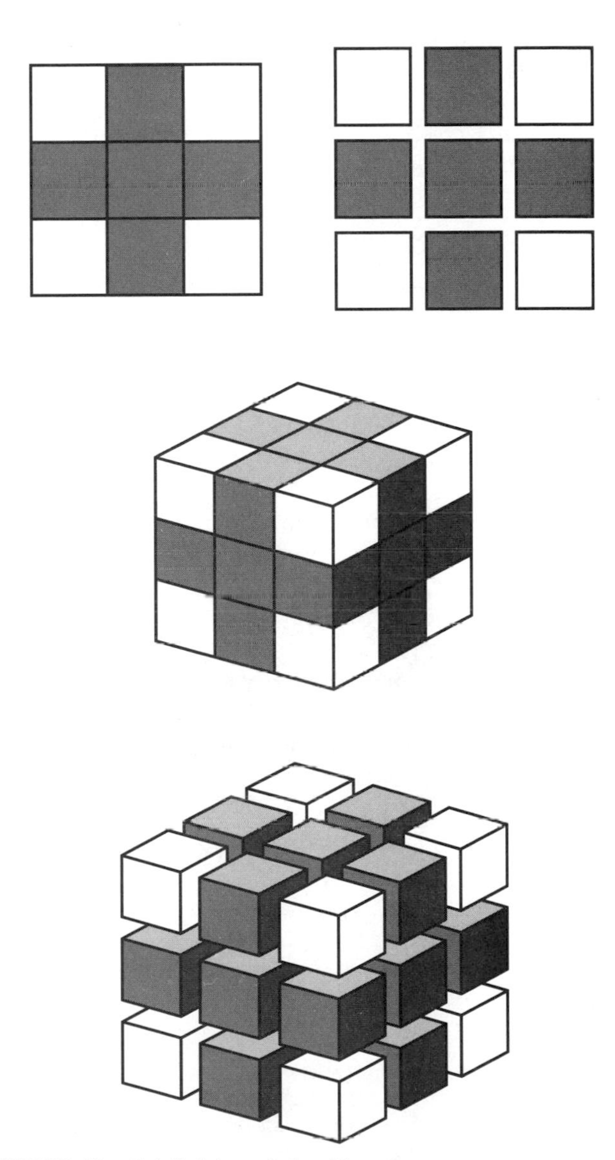

FIGURE 46. Subdivision of the sides of segments, squares, and cubes (and even hypercubes) into three equal parts yields 3, 9, 27, or 81 similar small objects—always a power of 3.

There are several ways to react to this observation. We can generate an additional row of the table to gain some additional information, but the conjecture is fairly firmly established with the five completed rows. We can observe that each entry is the sum of twice the entry directly above it plus the entry to the left of that one, so the sum of entries in one row is three times the sum of entries in the previous row—an argument that can easily be translated into a formal proof by mathematical induction. We may also use the explicit formula for the number of k-cubes in an n-cube, to sum a typical row:

$$\begin{aligned}
&\square(0,n)+\square(1,n)+\cdots+\square(n-1,n)+\square(n,n)\\
&=2^n+C(1,n)^{2n-1}+C(2,n)2^{n-2}+\cdots+C(n-1,n)2+C(n,n)\\
&=(2+1)^n=3^n.
\end{aligned}$$

All these approaches help explain why the rows sum to power of 3. But perhaps the most satisfying observation that justifies this fact is that we may divide the sides of an n-cube into three equal parts whose projections divide the entire cube into 3^n small cubes (Figure 46). The result is a small cube coming from each vertex of the original cube, one from each edge, one from each two-dimensional face, and so on. The final small cube is in the center. Thus the total number of small n-cubes, which is 3^n, is equal to the sum of the number of k-cubes in the n-cube—since there is one small n-cube for each point, edge, face, 3-cube, etc.

One of Friedrich Froebel's kindergarten gifts was a cube subdivided into 27 small cubes. He would have liked this final demonstration.

REFERENCES AND RECOMMENDED READING

1. Abbott, Edwin Abbott. *Flatland.* London, England: Seeley & Co., 1884; numerous reprintings, especially Blackwell's (1926) and Dover (1952).
2. Banchoff, Thomas. *Beyond the Third Dimension: Geometry, Computer Graphics, and Higher Dimensions.* New York, NY: Scientific American Library, W. H. Freeman & Co., 1990.
3. Banchoff, Thomas and Strauss, Charles. *The Hypercube: Projections and Slicing.* Chicago, IL: International Film Bureau, 1978.
4. Botermans, Jack. *Paper Capers.* New York, NY: Henry Holt and Company, 1986.
5. Barnsley, Michael. *Fractals Everywhere.* San Diego, CA: Academic Press, 1988.
6. Critchlow, Keith. *Order in Space.* New York, NY: Thames and Hudson, 1969.
7. Davidson, Patricia and Willcull, Robert. *Spatial Problem Solving with Paper Folding and Cutting.* New Rochelle, NY: Cuisenaire Company of America, 1984.
8. Dewdney, Alexander. *The Planiverse.* New York, NY: Poseidon Press, 1984.
9. Ernst, Bruno. *Adventures with Impossible Figures.* Norfolk, England: Tarquin Publications, 1986.

10. Heinlein, Robert. "...and He Built a Crooked House." In Fadiman, Clifton (Ed.): *Fantasia Mathematica.* New York, NY: Simon & Schuster, 1958.
11. Froebel, Friedrich. *Education by Development.* New York, NY: D. Appleton & Company, 1899.
12. Gardner, Martin. *Mathematical Carnival.* New York, NY: Alfred A. Knopf, 1975.
13. Hix, Kim. *Geo-Dynamics.* Conestoga, CA: Crystal Reflections, 1978.
14. L'Engle, Madeleine. *A Wrinkle in Time.* New York, NY: Farrar, Straus, and Giroux, 1962.
15. Manning, Henry Parker. *The Fourth Dimension Simply Explained.* New York, NY: Macmillan and Co., 1911.
16. Pearce, Peter and Pearce, Susan. *Polyhedra Primer.* Palo Alto, CA: Dale Seymour Publications, 1978.
17. Pearce, Peter. *Structure in Nature Is a Strategy for Design.* Cambridge, MA: MIT Press, 1978.
18. Peterson, Ivars. *The Mathematical Tourist.* New York, NY: W.H. Freeman & Co., 1988.
19. Rucker, Rudy. *The Fourth Dimension: Toward a Geometry of Higher Reality.* Boston, MA: Houghton Mifflin, 1984.
20. Tufte, Edward. *The Visual Display of Quantitative Data.* Cheshire, CT: Graphics Press, 1983.
21. Wells, David. *Hidden Connections, Double Meanings.* Cambridge, England: Cambridge University Press, 1988.
22. Wiebe, Edward. *Paradise of Childhood, Golden Jubilee Edition.* Milton Bradley (Ed.), including a "Life of Friedrich Froebel" by Henry Blake, Springfield, MA: Milton Bradley Company, 1910.
23. Winter, Mary Jane, et al. *Spatial Visualization.* Middle Grades Mathematics Project. Menlo Park, CA: Addison-Wesley, 1986.

Quantity

JAMES T. FEY

One of the principal factors in human intellectual development is our desire to make sense of the physical and biological worlds in which we live. We search historical records for clues that explain our present condition, and we devise theories that might predict the future. In nearly every description of the past or forecast of the future, prominent factors include quantitative attributes: length, area, and volume of rivers, land masses, and oceans; temperature, humidity, and pressure of our atmosphere; populations, distributions, and growth rates of species; motions of projectiles, tides, and planets; revenues, costs, and profits of economic activity; rhythms, intensity, and frequency of sounds, light, and earthquakes.

Perceptive observers have noted that patterns in objects can be modeled by numbers in ways that aid reasoning. It may be an exaggeration to say, as Lord Kelvin once claimed:[32]

> When you can measure what you are speaking about and express it in numbers, you know something about it; but when you cannot measure it, when you cannot express it in numbers, your knowledge is of a meager and unsatisfactory kind.

But it is not an exaggeration to say that the number systems of mathematics are indispensable tools for making sense of the world in which we live.

The human fascination with numbers is also reflected in countless examples of whimsical or superstitious numerology. From the Greek Pythagoreans to Martin Gardner's fictional Dr. Matrix,[10] people have found meaning—both sublime and sinister—in numerical values attached to letters, words, names, places, and dates. The endless variety of patterns in numbers has piqued the mathematical curiosity in millions of professional and amateur mathematicians of all ages. Unfortunately, those same patterns have served as the basis of various pseudoscientific enterprises—from astrology to numerology.

QUANTITY IN SCHOOL MATHEMATICS

Given the fundamental role of quantitative reasoning in applications of mathematics as well as the innate human attraction to numbers, it is not surprising that number concepts and skills form the core of school mathematics. In the earliest grades all children start on a mathematical path designed to develop computational procedures of arithmetic together with corresponding conceptual understanding that is required to solve quantitative problems and make informed decisions. Children learn many ways to describe quantitative data and relationships using numerical, graphic, and symbolic representations; to plan arithmetic and algebraic operations and to execute those plans using effective procedures; and to interpret quantitative information, to draw inferences, and to test the conclusions for reasonableness.

The skills required for these tasks are contained in the arithmetic of various number systems and in the generalizations of arithmetic reasoning to elementary algebra. The public recognizes these number systems by their common names (whole numbers, fractions, decimals); mathematicians use more formal terms (integers, rationals, real numbers). Regardless of their names, these number systems are well-known parts of mathematics and have been taught in school for centuries. Experienced teachers have devised countless clever strategies for developing student skill in solving traditional problem types. So it is entirely reasonable to ask, "What can be new and exciting about teaching quantitative reasoning?" Surprisingly, the answer ought to be, "Just about everything!"

Influence of Technology

School arithmetic and algebra have always been dominated by the goal of training students to manipulate numerical and algebraic symbols. The purpose of all this manipulation is to answer arithmetic problems or solve algebraic equations. The core of elementary and middle school mathematics features addition, subtraction, multiplication, and division of whole numbers and fractions; the core of secondary school

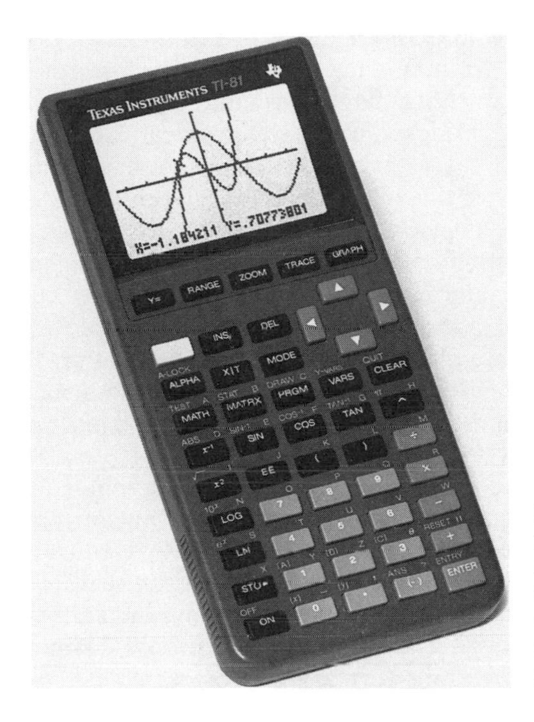

FIGURE 1. Hand-held calculators can now display graphs of all functions ordinarily studied in school mathematics. Some can even perform most common types of symbolic manipulation to simplify and solve equations.

mathematics covers similar operations on polynomial, rational, and exponential expressions.

In the past, proficiency with these routine manipulative skills has been a prerequisite for effective use of mathematics. However, the emergence of inexpensive electronic calculators and computers has changed that condition forever. It is now about 15 years since the technology of transistors, printed circuits, and silicon chips first made hand-held calculators available on the mass consumer market. Rapid progress in electronics has now produced solar-powered scientific calculators that perform arithmetic on numbers that can be entered and displayed in decimal, common fraction, or exponential form. Many calculators also have single-button subroutines for evaluating elementary functions and performing common statistical calculations. Programmable calculators offer more powerful capabilities, including graphing, symbolic manipulation, and matrix operations (see Figure 1). Each of these mathematical procedures is available in more powerful and sophisticated form through programs that run on desktop computers now widely available in schools.

The computational capabilities of machines—both existing and envisioned—suggest some exciting curricular possibilities. Elementary school students can now deal with realistic numerical data—very large and very small numbers in decimal and fractional form—without prerequisite mastery of the intricate computational algorithms for operations on those numbers. Middle school students can deal with questions about variables, functions, and relations expressed in algebraic language long before they master the rules for manipulating those expressions. In the world outside of school, almost everyone relies on calculators and computers for fast and accurate computation. But school curricula have yet to change significantly in response to these new conditions.

Calculators and computers are also having a profound effect on the nature of mathematics itself. With access to those tools, mathematicians can search for patterns in much the way that scientists explore results from experiments with systematically manipulated variables. The experimental mathematician can test special cases on a computer in a small fraction of the time required by "paper-and-pencil" algorithms. In many cases these calculations could not be done at all by traditional means, and the patterns that emerge would never have been seen. The experimental data of mathematics can be sorted, analyzed, and displayed graphically to reveal both regularities and variations. The ultimate standard for verification remains formal proof by reasoning from axiomatic foundations. However, calculators and computers have created a new balance between theorem-finding and theorem-proving.

Use of calculators and computers for mathematical work has also led to a dramatic increase in interest in algorithmic methods and results. Many of the deepest and most beautiful results of mathematics are those that guarantee the existence of numbers with interesting properties or solutions to important equations, yet those same theorems and their proofs quite often give no clue as to how one might effectively construct the promised object. Mathematical contemporaries of Euclid could prove that there is no largest prime number and that any natural number whatever can be factored uniquely into a product of primes. But mathematicians working today still devote great energy to practical and theoretical problems posed by the need to construct large primes and to find the promised factorizations of large composite numbers. The search for effective and efficient algorithms that will guide computer procedures has become a central aspect of both pure and applied mathematical research in our technology-intensive world.

Influence of Applications

A second fundamental change affecting school curricula is the extension of quantitative methods to nearly every aspect of contemporary personal and professional life. Although numbers have always been useful, their uses have been rather predictable and limited to well-defined familiar problems. Today, quantitative literacy requires an ability to interpret numbers used to describe random as well as deterministic phenomena, to reason with complex sets of interrelated variables, and to devise and interpret critically methods for quantifying phenomena where no standard models exist. Examples are all around us:

- U.S. census figures are used to describe our current population and to apportion resources to various social programs. How can the population and its characteristics best be counted?
- Several hurricanes strike Central and North America each fall. How can the "size" of each be measured in the most meaningful way?
- The consumer price index is used to calculate cost-of-living increases in Social Security payments and a number of other salary scales. How can inflation best be measured?
- Players on football teams in different conferences are often compared statistically to see who is best, in part to determine fair compensation. What data should be used to rank the quarterbacks most accurately?
- Banks, credit card companies, and airline and hotel reservation systems process billions of financial transactions daily, using national communication networks that are protected against errors and unauthorized intrusion. How can secure systems be devised and used intelligently?

Each of these problems and many others of similar complexity and significance require the ability to organize, manipulate, and interpret quantitative information. Skill in traditional written algorithms for arithmetic and algebra or in solution of traditional "types" of word problems is not only insufficient preparation for those tasks, it is largely irrelevant.

Quantitatively literate young people need a flexible ability to identify critical relations in novel situations, to express these relations in effective symbolic form, to use computing tools to process information, and to interpret the results of those calculations. The underlying mathematical ideas used in this modeling often extend beyond numbers and fractions to matrices, linear algebra, and the arithmetic of congruence classes. The useful computational tools extend beyond hand-held calculators to spreadsheets, data bases, and dynamic simulations.

Influence of Psychological Research

Another recent change in conditions for teaching about quantity in school mathematics is the emergence of an extensive body of research on human cognition. While there is a long history of research on mathematics teaching and learning from a psychological perspective, the past thirty years has seen an unprecedented search to identify the ways that young people develop understanding of number systems and their application. As a consequence, researchers are acquiring rich insight into the interplay between human cognitive development and the concepts, principles, and skills that we want young people to learn. This research shows real potential for informing decisions about design of curricula and instructional approaches in school mathematics.

FUNDAMENTAL CONCEPTS

The convergence of rapidly escalating demands for social and scientific application of quantitative skills with powerful new technologies that support those skills has prompted reconsideration of goals for school mathematics. To paraphrase the title of a 1982 report of the Conference Board of the Mathematical Sciences,[29] we are still asking, "What quantitative abilities will be fundamental in the future of mathematics?" Despite extensive professional debate over the past decade, there is as yet no consensus on a prudent course of change, and most evidence suggests that schools have not moved toward any radical change.

In a mature branch of mathematics such as number theory, analysis, or algebra, many fundamental concepts and operations can be presented in a coherent system of abstract ideas—a few definitions and axioms from which every other fact and principle follow logically. But this rigorous, efficient organization of contemporary mathematics is only the final product of an historical process in which fundamental ideas were used informally long before they become formal definitions and theorems. Furthermore, practical working knowledge requires more than an ability to recite or derive formal principles. It requires the ability to recognize quantitative relationships in a broad range of concrete situations as well as the technical skills to represent and reason about those relationships.

In thinking about school mathematics many mathematicians and teachers have argued that the best guide is a curriculum that retraces the meandering historical path by which numerical techniques have developed. Others suggest that we should capitalize on structural insights that have emerged at the end of that path, to provide for children a more efficient way to develop number concepts and techniques. There is little research evidence to suggest the right choice among these options, but it seems safe to say that quantitative understanding requires grasp of insights provided by each perspective. It seems important to convey to students, as quickly as possible, effective modern techniques for representing and reasoning about numerical data. But that instruction will undoubtedly be more successful if it is informed by understanding of the roots of numerical techniques in human experience and the path by which ideas and skills have evolved over time. Students must efficiently learn concepts, techniques, structural properties, and uses of the number systems[33]but with an honest portrayal of the many informal and halting ways that new mathematical ideas and methods actually develop.

Numbers and Operations

In searching for a framework of fundamental number concepts to be developed in school mathematics, it is helpful to begin with a simple question: How are numbers used? In common sources such as daily newspapers, cookbooks, instruction manuals, or household budgets, one will find a long list of situations in which numbers play a vital role. Furthermore, skill in quantitative reasoning is a critical prerequisite for success in any scientific, technical, or business occupation, and the list of ways that numbers are used in those fields is both long and diverse.

Designers of curricula are understandably frustrated by the challenge of selecting material that will prepare students for all problem-solving situations they might reasonably face outside school. However, a search for common features in quantitative reasoning tasks shows that they can be grouped into a few categories. One common analysis of number uses shows that every example involves one of three basic tasks:

1. MEASURING. To use operations of arithmetic to reason about size—to answer questions like "How many?" or "How much?"
2. ORDERING. To use numbers to indicate position in a sequence with the relations of "greater than" or "less than."
3. CODING. To provide identifying labels for objects in a collection.

Illustrations of these different tasks abound in ordinary life. Here are some particular examples:[7,33]

- Standard measurement tasks involving concepts such as length, area, volume, mass, and time all employ numbers to indicate size. The operations of addition, subtraction, multiplication, and division correspond directly to operations such as joining, comparing, or partitioning of objects that numbers measure. Other important concepts such as velocity, acceleration, and density also use numbers to indicate size, but they are usually derived by operations on basic number measurements.

- As customers enter a store they are often assigned numbers to indicate the order in which they will be served. Customers who enter early will have lower numbers than those who enter later—the order of arrival corresponds to the order of service numbers. In this case positive whole numbers are used to indicate order. It makes no sense to add or multiply service numbers, although subtraction might help to estimate expected waiting time.
- The teams in any athletic league are commonly listed in the order of their competitive standing, from first through last. However, without further information, those rankings tell little about the distance between teams in that order.
- In analyzing games of chance each possible outcome is assigned a number between 0 and 1 as its probability. Event A being more likely than event B corresponds to the probability $p(A)$ being greater than the probability $p(B)$. Furthermore, if A and B are disjoint, $p(A \cup B)$ should equal $p(A) + p(B)$. In this situation the assignment indicates a measure of likelihood. But those measures are then used to order events by likelihood. The operation of union for disjoint events corresponds to the addition of rational numbers.
- The uniforms of athletic teams generally have numbers for each individual player. While the numbers sometimes indicate an assigned position, arithmetic operations or relations involving those numbers seldom give any significant information. These numbers are used solely as labels.

This taxonomy of uses of numbers might seem too obvious to mention. But it offers the first step toward a framework for organizing the profusion of quantitative reasoning tasks into manageable families—a way to find significant themes among the details of number concepts, skills, and applications. With suitable refinement the taxonomy can help reveal to both teachers and students the experiential root meaning of numbers, to focus instruction on the forest as well as the trees.

For just that purpose Usiskin and Bell[33] have proposed a more detailed analysis of fundamental kinds of number uses. They suggest six different uses of single numbers:

- Counts for discrete collections (populations);
- Measures for continuous quantities (time, length, mass);
- Ratio comparisons (discounts, probabilities, map scales);
- Locations (temperature, time line, test scores);
- Codes (highway, telephone, product model numbers); and

- Derived formula constants (π in $A = \pi r^2$).

A parallel taxonomy suggests ways that operations on numbers can be matched to operations on objects that numbers describe:

- Addition models putting together or shifting;
- Subtraction models take-away, comparison, shift, or recovering an addend;
- Multiplication models size change, acting across, or use of a rate factor; and
- Division models ratios, rates, rate division, size change division, or recovering a factor.

While mathematicians and teachers might question the meaning of these categories and debate their completeness or independence, it seems certain that attention to such analyses will help focus instruction on the fundamental task of preparing students to use numbers effectively to solve problems. Examples of the different ways that numbers are used highlight the essential components in any quantitative reasoning task. In simplest form, quantitative reasoning involves phenomena, a number system, and a correspondence between phenomena and numbers that preserves essential structure. Each object is assigned a number in such a way that "similar" objects have "similar" numbers and relations among objects corresponding to relations in the number system. To understand this modeling process students need extensive experience with the structural properties of various kinds of number systems.

While students must certainly acquire comfortable skill in dealing with many specific uses of numbers, they also need to acquire a broader perspective on properties that number uses have in common. There is clear evidence from research in mathematics education that understanding fundamental structural properties of a mathematical system facilitates retention of the system and application to new situations. School mathematics should, therefore, emphasize the ways that different types of number systems serve as models of measuring, ordering, and coding, together with the ways that standard operations model fundamental actions in quantitative situations.

Variables and Relations

Elementary uses of numbers focus on descriptions and inferences concerning specific quantitative facts—the cost of 5 candy bars priced at 50¢ apiece, the area of a field that is 50 feet long and 30 feet wide, or the average speed of a car that travels 300 miles in 5 hours. Mastery of concepts required by such tasks is certainly a central and formidable task of school mathematics. However, for quantitative reasoning to yield results with greater power than unadorned number facts, it is essential that such reasoning be firmly rooted in general patterns of numbers and related computations.

The typical pattern is a relation among two or more varying quantities. For example,

- As time passes, the depth of water in a tidal pool increases and decreases in a periodic pattern.
- As bank savings rates increase, the interest earned on a fixed monthly deposit also increases.
- If a sequence of squares have sides 1, 2, 3, 4, 5, ..., the areas of those squares are 1, 4, 9, 16, 25, ...
- For any rectangle of base b and height h, the perimeter p is $2b + 2h$.

The key mathematical ideas required to reason about such patterns are the core concepts of elementary algebra: variables, functions, relations, equations, inequalities, and rates of change. In school mathematics today students spend a great deal of time working with variables as letter names for unknown numbers and with equations or inequalities that place conditions on those numbers. Algebra instruction focuses on formal procedures for transforming symbolic expressions and solving equations to find the hidden value of the variable.

But those skills are only a small part of the power that algebra provides. In each of the examples above, and in countless other similar problems, the conceptual heart of the matter is understanding relations among several quantities whose values change. The notion of variable that students must understand is not simply "a letter standing for a number" or "an unknown value in an equation." It must also include thinking about variables as measurable quantities that change as the situations in which they occur change.

Variables are not usually significant by themselves, but only in relation to other variables. In most realistic applications of algebra the fundamental reasoning task is not to find a value of x that satisfies one particular condition, but to analyze the relation between x and y "for all x." The most useful algebraic idea for thinking about relations of this sort is the concept of function.

To develop understanding required for effective application of algebra, students need to encounter and analyze a wide variety of situations structured by relations among variables. They need comfortable understanding of relational phrases such as "y depends on x," "y is a function of x," or "change in x causes change in y." It is helpful if they

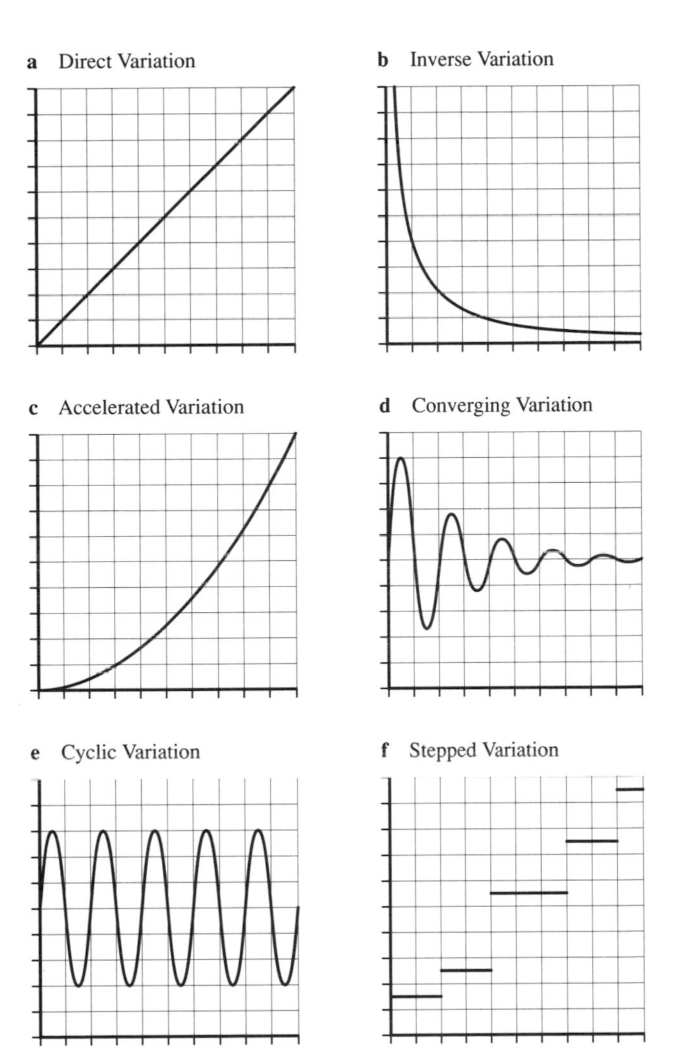

FIGURE 2. The behavior of fundamental types of relations among variables can be seen most readily from typical graphs. Graphs (a) and (b) illustrate direct and inverse relations, (c) and (d) show accelerated and converging variation, and (e) and (f) illustrate cyclic and stepped variation. Virtually all variation actually observed is a combination of these basic types.

develop a repertoire of criteria for characterizing and sorting, by structure, the relations they encounter. For instance, the report *Science for All Americans* of the American Association for the Advancement of Science[1] suggests that students should be sensitive to at least the following kinds of relations among variables (see Figure 2):

- Direct and inverse variation—as one variable increases, another also increases (or decreases) at a similar rate.
- Accelerated variation—as one variable increases uniformly, a second increases at an increasing rate.
- Converging variation—as one variable increases without limit, another approaches some limiting value.
- Cyclical variation—as one variable increases uniformly, the other increases and decreases in some repeating cycle.
- Stepped variation—as one variable increases, another changes in jumps.

The idea behind learning properties of whole families of relations is typical of all mathematics: recognition of structural similarities in apparently different situations allows application of successful reasoning methods to new problems. With the focus of algebra directed at variables and functions, equations and inequalities can be used to represent specific conditions:

- If the height of a projectile is a function of its time in flight with rule $h(t) = -16t^2 + 88t$, the equation $-16t^2 + 88t = 0$ asks when the projectile is at ground level (see Figure 3).
- If the population of a country (in millions) is a function of time with rule $p(t) = 120(2^{0.03t})$, the inequality $120(2^{0.03t)} \leq 200$ asks when the population will stay below 200 million (see Figure 4).

Of course, thinking about quantitative relations as functions encourages reasoning that extends beyond familiar equation-based questions to notions of rates of change, maxima and minima, and overall trends.

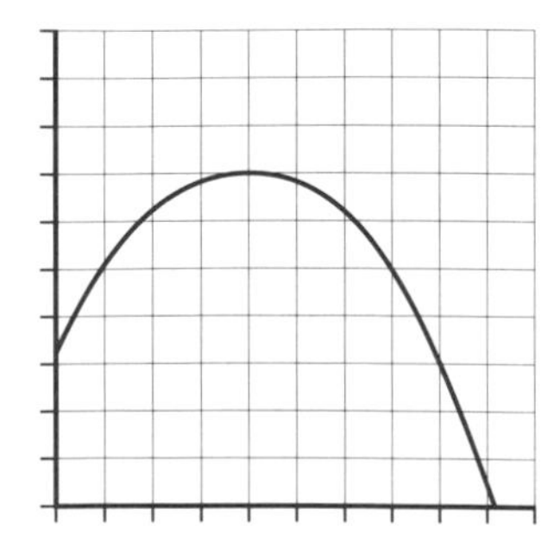

FIGURE 3. The standard parabolic trajectory becomes visible in a graph of the height of a projectile as a function of its time in flight.

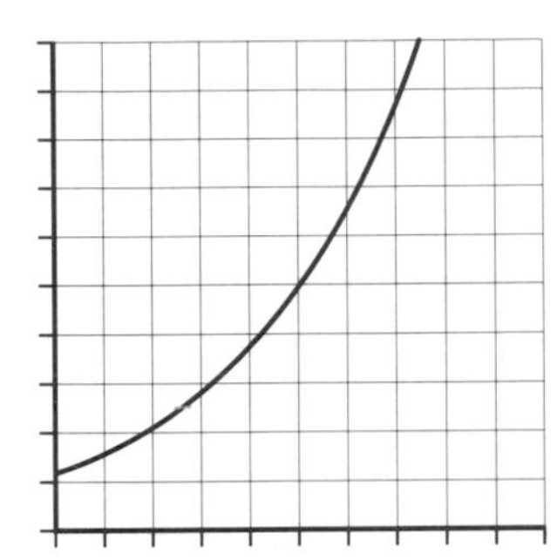

FIGURE 4. The common exponential curve represents the equation that describes the growth of a country's population.

While these questions are not generally considered central to school algebra, there can be no doubt that they are important considerations in any situation that algebraic expressions model.

PROCEDURES

The first step in effective problem solving is to analyze the problem to identify number concepts that match problem conditions. But that is only part of the modeling phase of solving problems—the conceptual description of what is known. Problem solving also requires inference of new information that gives new insight. In mathematics that inference invariably relies on systematic techniques for representing and manipulating information and, in quantitative problems, on procedures for calculating results. Recent analyses of mathematics pedagogy describe this kind of knowledge as *procedural* knowledge, in contrast to the *conceptual* knowledge required to identify fundamental ideas.[14] Procedural knowledge includes techniques required to represent information and to execute operations that yield solutions to specific numerical problems.

Numerical Representation

Formal mathematics is a subject that deals with mental constructs that are abstracted from patterns in objects. But mathematicians have also devoted a great deal of energy to find ways of representing ideas in concrete form. Their goal is a system of symbols that convey mathematical information effectively in unambiguous and compact form.

Representation of ideas serves as an aid to memory and as a medium for communication. In mathematics the representations become objects of study themselves—sources of new abstractions that, surprisingly often, serve as useful models of unanticipated patterns in concrete situations.

The fundamental idea that enables efficient representation of numbers is the place value system of numeration. Every whole number has a unique representation in the standard base 10 numeration system, and rational numbers can be expressed using decimal fractions or as quotients of whole numbers. These customary systems are sometimes replaced by place value systems with different bases, especially in cases where the alternative base has obvious advantages for a particular purpose.

While the place value system is taken for granted today, in thinking about mathematics teaching it is worth remembering that the evolution of such a powerful representation scheme took a very long time. There are signs in the record of early Mesopotamian mathematics that a base 60 numeration system, using few number symbols, was understood and used. However, the place value concept eluded Greek mathematicians in their golden era. It was not until Hindu mathematicians of the eighth century saw how to use 0 (zero) as a place holder that the foundation of place value notation was secured.

The second major task in representing numerical information is to express relationships that are true for all numbers, for many numbers, or for certain unknown numbers. The fundamental mathematical concepts involved are variables, functions, and relations. We now routinely use letters to name variables and to write rules for functions and relations. But again, it is worth recalling that the historical development of contemporary algebraic notation is a long story—testimony to the fact that the use of literal variables with algebraic syntax such as $y = x^3 - 4(x + 2)^{-1}$ is anything but obvious.

Graphical Representation

While traditional place value numerals and algebraic expressions are the most important symbolic forms for recording quantitative information, many other representational forms are in common use. The most popular are those that identify numbers with points in a geometric line or pairs of numbers with points in the plane.

For example, conditions on variables such as $|x - 2| \leq 3$ are quite common in algebra and its applications. The solutions can be given in a similar symbolic form, but it has become almost as common to display the results on a number line graph (Figure 5). Although this representation is certainly not as compact or computationally useful as the symbolic version, it conveys quickly a total picture of the quantitative condition.

The use of visual representation to display a relation among quantitative variables is especially effective when one variable is a function

FIGURE 5. Intervals portrayed on a number line provide an effective picture of the points that satisfy $|x - 2| \leq 3$.

of another. Here's a common example: The position of a piston with 4-inch stroke in an engine running at 3000 rpm is given by the function $y = 2\sin(100\pi t)$, where t is time measured in seconds. The pattern of piston positions is well displayed by a function graph (Figure 6). Like the number line graph, this visual image of a relation between two variables is not particularly effective as a computational aid, but it does convey the significant periodic pattern in piston motion in a way that is far less apparent from the symbolic form.

The use of number lines and coordinate graphs is a very familiar mathematical technique. However, the advent of graphing calculators and computer software has made a dramatic impact on the ease of producing graphs and thus on their usefulness. It is now possible to produce graphs quickly and accurately both from formulas and numerical data drawn from scientific experiments or from large data bases that computers have made accessible. As a result, graphic displays are becoming common and increasingly sophisticated. Thus it is important for mathematics students to become adept at interpreting graphic representations intelligently and to understand the connections among symbolic, graphic, and numerical forms of the same ideas.

There has been great optimism about the potential payoff of using these linked multiple representations as an aid in teaching. However, early experiments have revealed the fact that the messages provided by graphs are not grasped by young learners as easily as might be expected, while the effects of scale and the limited viewing window inherent in computer displays create surprising perceptual misconceptions.

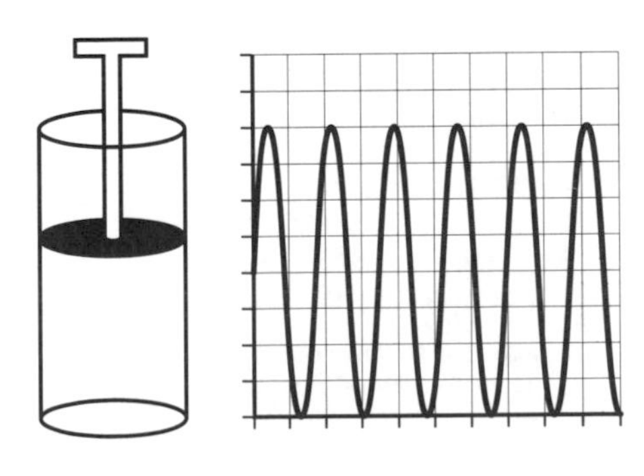

FIGURE 6. The motion of a piston is pictured by a sine graph, which conveys certain kinds of information more effectively than standard algebraic formulas.

Computer Representation

Cartesian graphs of numerical and algebraic patterns are only the most familiar strategies in an impressive array of visual representations for quantitative data. The burgeoning theory of graphs and networks includes many new techniques for representing situations with interacting quantitative and spatial structure. In some cases, network diagrams are used to display quantitative information like the costs of shipping foods or laying utility lines along various possible paths. In others, numerical representations such as matrices are used to organize and display geometric information like the number of possible paths between nodes of a graph. The field of exploratory data analysis includes many other new and effective techniques for representing numerical information in ways that convey meaning quickly, concisely, and effectively. The use of computers to produce those displays is becoming standard practice in all areas of applied mathematics.

One of the principal reasons for using compact symbolic forms to express relations among quantitative variables is the marvelous economy of capturing the full pattern of many numbers or n-tuples with a single symbolic sentence. However the abstraction required to reduce collections of data to symbolic rules also makes the information in those data less accessible to many potential users. Fortunately, computer tools also make display and reasoning with large data sets easy.

For example, the difference equation $y_{n+1} = 1.01y_n - 445$, where $y_0 = 5000$, describes the balance of a $5000 loan at 12% interest that is being paid back in monthly payments of $445. For most people the

Payment	Interest	Principal	Balance
			$5000.00
$445.00	$50.00	$395.00	$4605.00
$445.00	$46.05	$398.95	$4206.05
$445.00	$42.06	$402.94	$3803.11
$445.00	$38.03	$406.97	$3396.14
$445.00	$33.96	$411.04	$2985.10
$445.00	$29.85	$415.15	$2569.95
$445.00	$25.70	$419.30	$2150.65
$445.00	$21.51	$423.49	$1727.16
$445.00	$17.27	$427.73	$1299.43
$445.00	$12.99	$432.01	$867.43
$445.00	$8.67	$436.33	$431.10
$445.00	$4.31	$440.69	($9.59)

FIGURE 7. Spreadsheet representation of the balance of a $5000 loan at 12% interest that is being paid back in monthly payments of $445.

actual pattern in the dollar value of that loan and the distribution of payments to principal and interest is more informatively displayed in a simple spreadsheet such as that shown in Figure 7.

Of course, construction of this spreadsheet requires some ability to express relations in the symbolic form that has become standard with spreadsheets. In this case the formulas repeat with changing indices as follows:

Payment	Interest	Principal	Balance
			= 5000.00
= 445	= 0.01*D2	= A3–B3	= D2–C3
= A3	= 0.01*D3	= A4–B4	= D3–C4

Computer-generated numerical representations of algebraic expressions are proving to be a very useful tool in practical problem solving. For instance, to prepare the previous example, we calculated the appropriate monthly payment by experimental successive approximation, not by using the more conventional formula. But these representations also serve as a bridge from the concrete world of arithmetic reasoning to the more abstract world of algebra and statements that begin "for all x ...." Furthermore, the web of related representations comes full circle when computer curve-fitting tools are used to find symbolic rules that fit patterns in collections of numerical data.

Algorithms

The second major aspect of procedural knowledge consists of techniques commonly referred to as *algorithms* for using mathematical information to solve problems. An algorithm is a "precisely-defined sequence of rules telling how to produce specified output information from given input information in a finite number of steps."[23]

Developing student skill in execution of mathematical algorithms has always dominated school curricula at both elementary and secondary levels. The most prominent algorithms have been procedures for adding, subtracting, multiplying, and dividing whole numbers, common fractions, and decimals, along with the parallel operations on polynomial and rational expressions in algebra. But those are only the most basic and familiar among a vast library of routine mathematical tools. Euclid's algorithm, for example, is only one of several common methods for finding the greatest common divisor of two integers; the Sieve of Eratosthenes is only one of many algorithms for identifying prime numbers; the quadratic formula is one of many algorithms for solving quadratic equations; and there are dozens of algorithms for solving systems of linear equations and inequalities.

Design and application of algorithms are obviously at the heart of mathematics. The power of mathematics comes from the way that its abstract ideas can be applied to solve problems in contexts with no surface similarities. The algorithms of arithmetic and algebra that are used in business and economics are the same as those used in physics and engineering. At the same time the context-independent nature of mathematical algorithms makes them easily programmed for computer execution. This fact has major implications for school curricula: any specific algorithm that is of such fundamental importance and broad applicability to merit inclusion in elementary or secondary school will certainly have been programmed and made available in standard calculator and computer software. Inexpensive calculators can perform most numerical, symbolic, and graphic algorithms that are taught in school. Thus, current technology seriously undermines any argument that students must develop proficiency in executing any particular algorithm because they will need that skill later in life.

At the same time that learning of specific algorithms has diminished in importance for school mathematics, it has become far more important for everyone doing quantitative work to have general understanding of the algorithmic point of view.[9,23,26] To be an intelligent user of computer-based algorithms, it is useful to understand such attributes as accuracy, economy, and robustness as well as fundamental mathematical concepts like induction and recursion that are too little appreciated in traditional curricula. In short, the algorithmic aspect of mathematics takes on a very different appearance when calculators and computers take over routine systematic procedures. This new condition requires fundamental reconsideration of goals for quantitative study in school mathematics.

Conceptual and Procedural Knowledge

Calculators and computers have clearly taken over routine aspects of both representation and manipulation of quantitative information—the two key components of procedural knowledge. The task of translating these new conditions into new goals for curricula poses a critical psychological question concerning the interplay between conceptual and procedural knowledge. Many mathematics educators worry that extensive use of calculator and computer tools, with corresponding de-emphasis of training in skills, will undermine development of conceptual understanding, proficiency in solving problems, and ability to learn new advanced mathematics.

The interaction of understanding and skill in mathematics has been studied and debated intensely for many years but with renewed enthusiasm in the past decade. A recent meta-analysis of over 70 research studies[13] concluded that wise use of calculators can enhance student conceptual understanding, problem solving, and attitudes toward mathematics without apparent harm to acquisition of traditional skills. More limited research in algebra suggests similar conclusions. While there is a great deal of work in progress on this issue, the principal reported results are from Demana and Leitzel.[8]

However, in almost all of those experiments the calculator or computer was used to complement instruction in traditional arithmetic and algebraic skills. What remains an open and very important problem is to determine the consequences of more daring experiments in which students are taught to rely more heavily on technological help with arithmetic and symbolic manipulation. It seems safe to say that the debate over proper consideration of conceptual and procedural knowledge will continue for some time. It is certainly *the* central issue raised by the impact of technology in school mathematics.

Number Sense

While there is considerable debate concerning the risks and benefits of shifting attention in school mathematics from traditional skills to concepts and problem solving, there is no disagreement about the importance of developing student achievement in a variety of informal aspects of quantitative reasoning, to develop what might be called *number sense.* Even if machines take over the bulk of computation, it remains important for users of those machines to plan correct operations and to interpret results intelligently. Planning calculations requires sound understanding of the meanings of operations—of the characteristics of actions that correspond to various arithmetic operations. Interpretation of results requires judgment about the likelihood that the machine output is correct or that an error may have been made in data entry, choice of operations, or machine performance. (Development of number sense is discussed in detail in the February 1989 issue of *The Arithmetic Teacher,* especially in the article by Howden.[16])

There are two fundamental kinds of skill required to test numerical results for reasonableness. First is a broad knowledge of quantities in the real world: Is the population of the United States closer to 20 million, 200 million, or 2 billion? Is the speed of an airplane closer to 100, 1000, or 10,000 kilometers per hour? What are approximate percent rates for a sales tax, a car loan, the tip at a restaurant, or success of a major league baseball hitter? While this sort of information isn't part of formal mathematics, it is an invaluable backdrop for judgment of arithmetic applied to real problems.

The second component of computational number sense is the ability to make quick order-of-magnitude approximations. As an electronic calculator produces an exact answer, it is important for users to check that the displayed results are "in the right ball park." This means, for instance, determining by quick rounding and mental arithmetic that $345 + 257 + 1254$ is approximately 1850 or that 85×2583 is approximately 200,000. Skillful mental calculation of this sort is *not* achieved by extensive training in mental execution of traditional written algorithms, but in flexible application of place value understanding and single-digit arithmetic—a very different agenda than the goals of traditional school arithmetic. Since there has been considerable attention given to informal arithmetic and computational estimation over the past decade, there now are clear goals, creative curriculum materials, and effective teaching suggestions for this important but long-neglected topic.

Symbol Sense

There is almost certainly a comparable informal skill required to deal effectively with symbolic expressions and algebraic operations—to cultivate student *symbol sense*—but ideas and instructional materials in this area are not as fully developed. A reasonable set of goals for teaching symbol sense would include at least the following basic themes:

- Ability to scan an algebraic expression to make rough estimates of the patterns that would emerge in numeric or graphic representation. For example, given $f(x) = 50 * 2^x$, a student with symbol sense could sketch the graph of this function and realize that function values will be positive and monotonically increasing—with small values of $f(x)$ for negative x and rapidly increasing values for positive x.
- Ability to make informed comparisons of orders of magnitude for functions with rules of the form n, n^2, n^3, ..., and k^n. This skill, a bridge between number and symbol sense, plays an important role in judging the computational complexity of algorithms for mathematical and information-processing tasks that are at the heart of computer science.
- Ability to scan a table of function values or a graph or to interpret verbally stated conditions, to identify the likely form of an algebraic rule that expresses the appropriate pattern. For example, given the following table, a student with symbol sense could predict that the rule for the best-fitting function is likely to be of the form $f(x) = mx + b$ with m approximately 15 and b about 500:

Sales x	0	10	20	30	40	50
Costs $f(x)$	510	675	825	960	1100	1240

- Ability to inspect algebraic operations and predict the form of the result or, as in arithmetic estimation, to inspect the result and judge the likelihood that it has been performed correctly. For instance, a student should realize almost without thinking that the product of linear and quadratic polynomials will be a cubic polynomial.

- Ability to determine which of several equivalent forms might be most appropriate for answering particular questions. For instance, good symbol sense should allow students to realize that the factored form of a polynomial readily yields information about its zeroes but makes very difficult calculation of derivatives or integrals.

Promising work from current projects shows how numerical and graphic computer tools can be used effectively to build student intuition about algebraic symbolic forms. Nevertheless, the development of more general symbol sense remains an important research task on the path to new approaches for developing conceptual and procedural knowledge of quantity.

NUMBER SYSTEMS

For a great many students mathematics is a vast, loosely connected collection of facts, procedures, and routine word problems. However, it is important to remember that the unique power of mathematical concepts depends on abstract meaning, which lies at the heart of any specific embodiment. Learning the fundamentals of any branch of mathematics should include recognition of those deep structural principles that determine the relations among its concepts and methods. For number systems a rather small collection of big and powerful ideas determine the structure of each system. When one steps back from specific details, it becomes clear that a few central principles govern all algebraic and topological properties of numbers. These principles can be used to derive all specific facts of various number systems and to guide the match between formal systems and significant quantitative problems.

In the historical development of number systems, the progression began with the natural numbers. Extensions over many centuries added fractions, then negative numbers, and, finally, a rigorous characterization of real numbers. From a perspective near the end of the twentieth century it is possible to organize all those structures from the top down:

- The real number system R is the only complete ordered field.
- The rational number system Q is the smallest subfield of R.
- The integer number system I is the smallest ring in R that includes the multiplicative identity.
- The natural number system N is the smallest subset of R that includes the multiplicative identity and is closed under addition.

In the terse form that is characteristic of formal mathematics, these four statements contain a great deal of information about structure. They imply that each number system is a set with two binary operations and a binary order relation; that the operations are commutative and associative; that multiplication distributes over addition; that there are two identity elements, one for addition and the other for multiplication; and that the operations interact with the order relation in familiar ways.

There are, however, other important properties of the individual number systems that are not so apparent from such minimalist characterizations. There are significant differences in the algebraic and topological properties of the various systems, differences that make each of special interest from both pure and applied perspectives. Analysis of those differences, in progression from the simplest to the most subtle, helps develop student insight into the nature of numbers and number systems. While students should emerge from school mathematics with rich conceptual and procedural knowledge, it is also important that they have some sense of the theoretical principles that provide logical coherence to number systems.

Natural Numbers and Integers

The fundamental additive, multiplicative, and order structures of the natural numbers and integers are based on several simple but powerful principles. First is the principle of finite induction:

> If M is a set of natural numbers that contains 1, and if M contains the number $k+1$ whenever it contains the number k, then M contains all the natural numbers.

This property implies that the natural numbers (and their extension to all integers) form a discrete set, a sequence of equally spaced elements with no number between any integer k and its successor $k + 1$. They provide a set of tags for ordering stages in any process that can be viewed as occurring in a sequence of discrete steps.

The finite induction principle is used to define concepts with integer parameters, like x^n, and to prove propositions that involve all natural numbers. For example, to prove that $1 + 3 + 5 + \ldots + (2n - 1) = n^2$ for all n, one depends on the principle of finite induction:

1. Let M be the set of numbers for which the equation is true. Since $1 = 1^2$, we know that $1 \in M$.
2. Now suppose that $k \in M$. Then $1+3+5+\ldots+(2k-1) = k^2$. It follows that $1 + 3 + 5 + \ldots + (2k - 1) + (2k + 1) = k^2 + 2k + 1 = (k + 1)^2$, so the equation is also true for $k + 1$. Hence $k + 1$ also belongs to M.
3. It follows from the principle of finite induction that M contains all the natural numbers, so the formula must be true for all n.

The method of proof by mathematical induction is used throughout mathematics, providing special power in combinatorial propositions like the binomial theorem. It has become particularly important as a proof technique in computer science, where discrete algorithmic processes are the central objects of study.

While natural numbers and integers share the discrete order structure implied by the principle of induction, there is one critical difference between the two systems—the existence of additive inverses for integers: for every integer a there is an integer $-a$ such that $a + -a = 0$. This makes the integers into an additive group, implies that subtraction is defined for all ordered pairs of integers, and shows that every equation of the form $a + x = b$ has a unique solution in I.

Although the additive structures of N and I are extremely regular and easy to work with, multiplication and division of natural numbers and integers hold much more interesting challenges. Since the integers contain no multiplicative inverses (except the trivial cases of 1 and -1), division is a restricted operation in N and I, and many equations of the form $ax = b$ have no integer solutions. Furthermore, there is no simple pattern suggesting which multiplication equations (or related divisions) are solvable. The integer 24 is divisible by 2, 3, 4, 6, 8, and 12, but its neighbor 23 has no proper factors and 25 has only one proper factor. A set of 24 objects can be partitioned into equal subsets in six different ways, but a set of 23 cannot be partitioned in any such way.

Multiplication and division of integers are governed by two principal properties. The fundamental theorem of arithmetic guarantees that any positive integer can be written as a product of prime factors in exactly one way. The division algorithm guarantees that for any positive integers a and b there are unique integers q and r such that $a = bq + r$ with $0 \leq r < b$. These two principles are of enormous practical and theoretical significance in the theory of numbers and, in more general form, in algebra.

The first—the prime factorization theorem—is one of many similar results in mathematics showing how complex expressions can be studied effectively when they are written as a combination of irreducible factors. These applications range from the mundane task of finding least common multiples or greatest common factors to the parallel fundamental theorem of polynomial algebra which assures that any polynomial with complex coefficients can be written as a product of linear factors (from which the zeroes can be easily obtained).

The division algorithm is, of course, basic to the familiar procedure for long division of natural numbers and decimals as well as to the parallel factor theorem of polynomial algebra. It provides the essential concept for developing the arithmetic of congruences: For any integers a and b, $a = b \pmod{m}$ if and only if $a = mk+b$ for some k. The finite cyclic groups and fields that arise from this theory have proven useful in dramatic ways as models for discrete phenomena, including increasingly important applications in computer science, in cryptography, and in transmission and storage of business and governmental information.

Rational Numbers

The smallest number system that includes elements representing each possible division of integers a/b (for nonzero b) is, of course, the rational number system Q. Mathematicians call Q a *field,* a term used to describe other structures with similar number-like properties. In Q every nonzero element has a multiplicative inverse, and every linear equation of the form $rx + s = t$ has a unique solution for rational r, s, and t (for nonzero r). However, this algebraic power is gained at the expense of simplicity.

The standard ordering of rational numbers makes them a *dense* set—between any two rational numbers there is a third rational number. In particular, there are positive rational numbers as small as one might wish. On the other hand, for any rational numbers a and b, there is an integer n such that $na > b$; this property makes the rational numbers into an *Archimedean* ordered field. While the operations and ordering of rational numbers are significantly more complex than integers, the density and Archimedean properties of Q combine to lay the groundwork for precision in measurement, guaranteeing that a unit of any desired refinement can be used to cover a length of any finite extent.

Real Numbers

The natural numbers, integers, and rational numbers provide formal systems to model the structures of many practical quantitative reasoning tasks. But unresolved questions raised as long as 2000 years

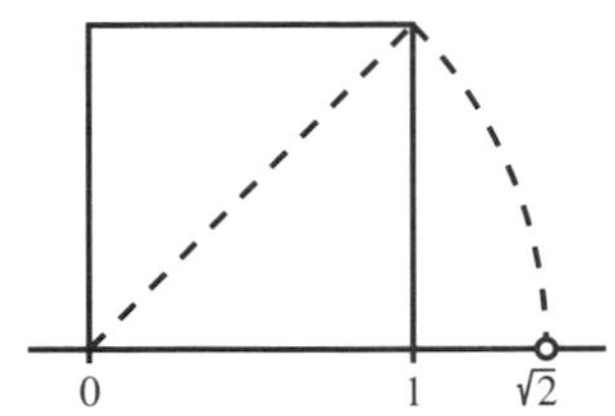

FIGURE 8. The position on a rational number line corresponding to the length of the hypotenuse of a right triangle with legs of length 1 has a hole, since there is no rational number equal to $\sqrt{2}$.

ago make it quite clear that the rational numbers are not the last word in number systems. The proof that there is no rational number whose square is 2 (or 3, or 5, or any other integer that is not a perfect square) reveals an algebraic incompleteness in the rational number system (see Figure 8). When numbers are used as measures of geometric figures, the Pythagorean theorem shows that there are line segments with no rational measures. There are "holes" (although not very big holes) in a number line that has only rational coordinate points.

The rational numbers can be extended in a variety of ways to include elements that fill some of these holes and that fulfill specific algebraic or geometric needs. The extension $Q(\sqrt{2}) = \{a + b\sqrt{2}: a, b \in Q\}$, for instance, is an ordered field, under a suitable definition of addition, multiplication, and inverses. However, the only *complete* ordered field—one that fills *all* the holes—is the real number system R. It is an ordered field in which every nonempty subset that is bounded from above has a least upper bound in R. A key theorem of number systems, one that establishes a distinctive role for R, is that any such complete ordered field must be isomorphic to R.

Since the real numbers seem only to fill "infinitesimal" holes on the rational number line, several other differences between the two number fields are genuinely surprising. First, while every rational number is the solution of a simple equation $ax = b$ where a and b are integers, there are transcendental real numbers (like e and π) that are *not* solutions of any such polynomial equation. Furthermore, while the rational numbers can be placed in one-to-one correspondence with the natural numbers and are thus countably infinite (a surprising result that was not comfortably understood until early in this century), this is not true for the real numbers. In fact, the transcendental numbers alone are more numerous than the algebraic numbers—those that arise as solutions to rational algebraic equations. While this last result was proven at least 100 years ago through very clever reasoning with transfinite cardinal numbers, there are still subtle outstanding questions about the character of specific real numbers.

The real numbers provide a significant step in the development of quantitative concepts and methods in another fundamental sense. While the natural numbers, integers, and rational numbers are each infinite sets of numbers, their primary use is to count, order, and compare finite sets of discrete objects. The real numbers provide the essential mathematical tool to describe and reason about infinite and infinitesimal processes. They alone support rigorous development of the concepts of limit and continuity; they provide the bridge to analysis of motion and change.

Complex Numbers

The extension from rational to real numbers enables solution of many simple and significant algebraic equations. But it leaves an equally significant collection of algebraic equations still unsolvable. Simple polynomial equations like $x^2+1=0$ or $x^2+x+1=0$ have no real roots. The number system required to give meaningful solutions to these equations, and to all polynomial equations in general, is the complex numbers C. The complex numbers constitute the smallest possible field extension of the real numbers that contains an element i with square equal to -1, the required root of $x^2+1=0$. Remarkably, the extension to deal with this single equation provides solutions to all other polynomial equations and opens a rich structure of mathematical properties and applications.

Every complex number can be expressed in the form $a+bi$, where a and b are real numbers. Thus the complex numbers are determined by ordered pairs of real numbers. While the real numbers can be ordered in one-to-one correspondence with the points of a line, the complex numbers correspond to points of a two-dimensional plane and are not linearly ordered. This loss of simple order might seem to promise a much more complicated life in C than in the real numbers or their subsets. However, it brings along benefits as well. The correspondence between complex numbers and points in the plane opens a powerful connection betweeen the arithmetic and algebra of C and the geometry of shapes and transformations in the plane (see Figure 9).

The complex numbers include some numbers originally described as "imaginary" by mathematicians who could not admit the possibility of a negative square. Nevertheless, they have proven useful as models of many very real physical phenomena, from the flow of alternating

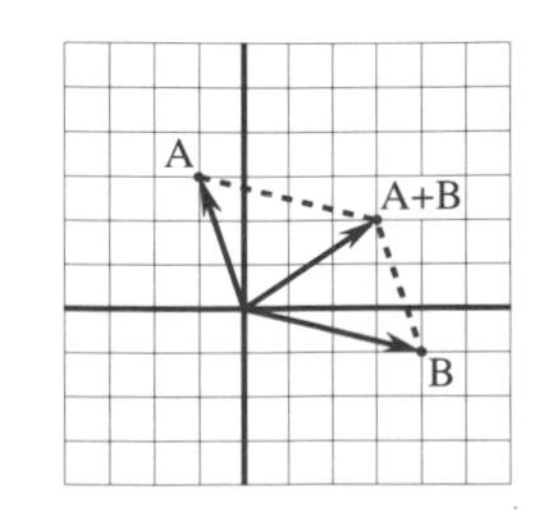

FIGURE 9. Points in the plane correspond to complex numbers, with addition of vectors in the plane reflecting addition of complex numbers. Multiplication is more complicated—the magnitudes of the vectors multiply as expected, but the angles add.

electrical current to the flow of air over an airplane wing. They also settle a fundamental algebraic question of pure mathematics: every polynomial of degree n has exactly n linear factors. Thus every polynomial equation has at least one and at most n distinct complex roots.

New Number Systems

Our sketch of fundamental principles of number systems covers very familiar ground. When mathematicians of the late nineteenth century showed that the real number system is the unique complete ordered field, following earlier proofs by Gauss and others that the complex number system is algebraically closed, it seemed that the story of number systems was complete. While that is, in some sense, an accurate statement, the development of new number systems is by no means finished.

For example, since their invention in the mid-nineteenth century, the algebra of matrices has become an invaluable tool for reasoning about complex numerical data. A matrix is a kind of super-number; within certain families of matrices, the operations of addition and multiplication have algebraic properties very similar to those of the real numbers. The most prominent exception is the fact that matrix multiplication is noncommutative—a fact that has many important consequences in the theory of linear algebra. Matrices are particularly useful for describing complex sets of quantitative data such as those that computers routinely manage.

The application of computing to quantitative reasoning has stimulated development of mathematical systems in another direction of both practical and theoretical interest. Despite their seemingly endless memory and instantaneous speed, computers work not with the familiar number systems such as I, Q, or R, but in finite approximations of those systems whose faithfulness is limited by the ability of computer languages to represent numbers with only a finite number of positional places. These "truncated" models of number systems do not obey the conventional structural properties of numbers (such as associativity of addition). Thus it seems important that students extend their study to include the structural properties of those finite systems that underlie so much of their actual quantitative work.

The discovery of number-like mathematical systems like matrices that fail to obey structural properties that our naive intuition tells us are true was a dramatic step in the development of modern mathematics. Contemporary algebra originated in an attempt to provide a theory to explain the structural properties of various number systems. In the last 150 years algebra has generated a rich array of abstract theories that spring from study of structure inherent in various operations and relations on sets. Mathematicians have shown that generalization of number systems can provide a stimulating intellectual playground. But they have also shown that this abstract mathematical realm frequently has impressive practical applications. Although groups, rings, fields, lattices, boolean algebras, monoids, and Turing machines were created primarily as abstract possibilities, they are now used routinely as tools for research on fundamental problems of computing and information sciences.

During the middle of this century, mathematics was strongly influenced by interest in exploring generalizations of number systems. In 1973 Garrett Birkhoff[3] wrote that, "by 1960 most younger mathematicians had come to believe that all mathematics should be developed axiomatically from the notions of set and function." Furthermore, he and MacLane, "wrote another 'Algebra' which went further in the direction of abstraction, by organizing much of pure algebra around the central concepts of morphism, category, and 'universality'." Innovative school mathematics programs of the 1960s explored the possibilities of organizing curricula around similar abstract structural concepts.

Fashions change, in mathematics as well as in design of human artifacts. Today the abstract axiomatic point of view seems much less promising as a guide to either mathematics research or education. Nonetheless, there *are* central principles that lie at the heart of number systems and algebra. They provide coherent organization for what can be an impenetrable maze of specific facts and techniques, and this organization is as useful for students as for practicing mathematicians. Thus it seems wise for curriculum planners to identify and build from such principles as they plan school curricula.

APPLICATIONS

School mathematics must develop in students an understanding of basic principles, proficiency in techniques, and facility in reasoning. But the ultimate test of school mathematics is whether it enables students to apply their knowledge to solve important quantitative problems. The ability to solve problems is not only the most important goal of school mathematics but also the most difficult educational task.

The term "word problem" strikes terror in the hearts of mathematics students of all ages. The key first step in effective work on problems is to identify in problem situations concepts that are structurally similar to those of number systems. Traditional approaches to this task can be sorted into two broad classes. The pragmatic approach helps students cope with a variety of classical (and nearly routine) problem types. The aim is to provide students with strategic guidelines for each problem type—a special chart for organizing information about time/rate/distance problems, a dictionary for translating key quantitative words into symbolic expressions, and so on. The more ambitious approach attempts to train students to use generic high-level strategies (or heuristics) that apply to problems in many different areas.

It seems fair to say that neither approach is demonstrably effective in providing students with confident and transferable modeling and problem-solving skills. The (unfortunately) popular "key words" approach fails because the flexible, versatile, and often ambiguous structure of ordinary language cannot be translated into mathematical statements by any dependable algorithm. At the other extreme, while students can learn generic high-level heuristics suggested by Pólya and others, it has proven very difficult to develop their facility in the kind of metacognitive monitoring of thought that is required to deploy those heuristics effectively in specific situations. Recent work to develop a metacognitive perspective on problem-solving strategies shows promising but not yet definitive results.[30]

Modeling

While the search continues for effective new strategies to teach problem solving, there is an equally significant change emerging in thinking about the nature of quantitative problems themselves. In many contemporary applications of mathematics one thinks less about solving specific well-defined problems and concentrates instead on constructing and analyzing mathematical models of the problem setting. The classical quantitative problems of school mathematics usually include numerical information and a single question that can be answered by a numerical calculation or by solving an equation. Outside school, problem situations generally have missing or extraneous information as well as many ill-defined questions.

In a mathematical modeling approach, the first step is to identify relevant variables. The next is to describe, in suitable formal language, relations that represent cause-and-effect connections among those variables. Specific questions can then be posed in terms of input or output values or global properties of the modeling relations. Finally, computer tools can be used to answer those questions by numerical, graphic, or symbol methods.

Measurement

The most common sources of numerical variables are measurements. Thus the theory and technique of measurement play important roles in quantitative literacy. Like the arithmetic of number systems, measurement feels like a familiar and well-known facet of school mathematics—hardly in need of new thinking. However, this critical interface between mathematics and its applications is not a remarkably successful topic in the curriculum.

The prototypical measurement tasks in school mathematics are finding length, area, and volume of geometric figures. It seems fair to say that for most students learning about measurement includes brief exposure to a few standard units for length and then practice in use of formulas for perimeter, area, and volume based on those length units. Area is [*length* × *width*] or [(1/2) × *base* × *height*] or [$\pi \times r^2$], volume is [*length* × *width* × *height*] or [$\pi \times r^2 \times h$], and so on. Most exercises become arithmetic practice in the formula of the lesson at hand.

Students exposed to this formal approach to measurement generally form limited and very rigid conceptions of length, area, and volume. Confusion of area and perimeter is a depressingly common error on student assessments. The common "rule" followed by unthinking students, regardless of any wording in the problem statement, is that if there are two numbers attached to sides of a pictured rectangle one multiplies them; if there are numbers on each side of a rectangle, one adds them.

The emphasis on formulas also leaves students ill prepared to deal intelligently with the approximate nature of real measurements, the cumulative effects of errors in combinations of measurements, and the generalization to irregular shapes that occur in so many practical applications or to the curves and surfaces that are fundamental in calculus. Furthermore, few students realize or take advantage of structural similarities that underlie most applications of measurement.

At the heart of any measurement process is a mapping that assigns numbers to objects. The mapping assigns measure 1 to some designated unit. Other objects are then covered by copies of the unit. The choice of unit element is arbitrary, but once the choice is made, it provides the standard by which all others are measured. Thus every measurement consists of a unit and a number—the number of whole and partial copies of the unit needed to exactly cover the measured object. The mathematics student who understands this principle—as a general property of many important measurements—has acquired productive insight into the connection between real situations and quantitative models.

The unit and covering properties of measurement explain quite clearly just what is being indicated by any particular measurement; moreover, the attachment of units to measurements can be exploited to guide formal reasoning about scientific principles. In many sciences quantitative reasoning is guided by a well-defined algebra of quantities commonly called "dimensional analysis." In this method each arithmetic operation is performed not only on numbers but on the units as well. If the end result is a number whose units are appropriate for the problem, the dimensional analysis lends support to the appropriateness of the operations that have been performed. While this attention to units as well as numbers in measurement is not as common in mathematics as in science, it has strong supporters among those who have been concerned with helping students make the connection between formal mathematics and its applications.[18,21,31]

The theory and practice of measuring quantitative concepts in the physical world have a long history in mathematics and its teaching. However, just as many classical mathematical methods have been generalized and applied in new domains, measurement has been extended to important uses throughout the social sciences. While the basic idea is the same—assigning numbers to objects or events—these new measures often obey structural properties that are very different from the measures of length, area, and volume.

Political scientists and sociologists have designed a variety of measures of influence or power in social situations. Economists have devised measures of costs and benefits to quantify options in decision making. Psychologists and educators use a vast array of measures to describe aptitudes and achievements of individuals. Statisticians measure probable cause-and-effect relations among many different kinds of stochastic variables. In each case, numbers, operations, and relations are used to model significant structural properties of situations. Sometimes classical principles and concepts are directly applicable. But it is increasingly common that effective quantitative reasoning in the social and human sciences requires understanding of aspects of number that permit flexible construction of new responses to new situations.

GOALS

Without question the most important goal of school mathematics is to develop students' ability to reason intelligently with quantitative information. The mathematical concepts, techniques, and principles that model quantitative aspects of experience are provided by structures of number systems, algebra, and measurement that have long been the heart of school curricula. However, the emergence of electronic calculators and computers as powerful tools for representing and manipulating quantitative information has challenged traditional priorities for instruction in those subjects. It no longer makes sense to devote large portions of the school curriculum to training students in arithmetic or algebraic algorithms that can be performed quickly and accurately by low-cost and convenient calculators. The availability of powerful aids to computation has also led to a dramatic increase in the range of situations to which quantitative reasoning is being applied. Thus school mathematics must prepare students to use their knowledge of number, algebra, and measurement in flexible and creative ways—not only in routine, predictable calculations.

To prepare students for the challenge of quantitative reasoning in the modern world, school mathematics must develop students' abilities to

- Understand fundamental properties of number systems and the match between those mathematical systems and real-life situations in which they are embodied.
- Describe and interpret quantitative structures using symbolic, verbal, and graphic representations.
- Perform both exact and approximate calculations involving arithmetic and algebraic ideas by various appropriate methods—mental operations, paper-and-pencil techniques, calculators, or computers.
- Apply numerical and algebraic expertise to solve both routine and original quantitative problems.

The school experience likely to develop these general skills and understandings must be rich in opportunities to explore interesting and important quantitative situations as well as in the structures that illuminate mathematical ideas embodied in specific settings.

REFERENCES AND RECOMMENDED READING

1. American Association for the Advancement of Science. *Science for All Americans.* Washington, DC: American Association for the Advancement of Science, 1989.
2. Baumgart, John K. "The history of algebra." In Hallerberg, Arthur E. (Ed.): *Historical Topics for the Mathematics Classroom: Thirty-first Yearbook of NCTM.* Washington, DC: National Council of Teachers of Mathematics, 1969, 232–260.
3. Birkhoff, Garrett. "Current trends in algebra." *American Mathematical Monthly,* 80 (1973), 760–782.
4. Brainerd, Charles J. *The Origins of the Number Concept.* New York, NY: Praeger, 1979.
5. Dantzig, Tobias. *Number: The Language of Science, Fourth Edition.* New York, NY: Macmillan, 1959.
6. Davis, Harold T. "The history of computation." In Hallerberg, Arthur E. (Ed.): *Historical Topics for the Mathematics Classroom: Thirty-first Yearbook of NCTM.* Washington, DC: National Council of Teachers of Mathematics, 1969, 87–117.
7. Davis, Philip. *The Lore of Large Numbers.* Washington, DC: Mathematical Association of America, 1961.

8. Demana, Frank and Leitzel, Joan. "Establishing fundamental concepts through numerical problem solving." In Coxford, Arthur F. and Shulte, Albert P. (Eds.): *The Ideas of Algebra, K–12: 1988 Yearbook of the NCTM.* Reston, VA: National Council of Teachers of Mathematics, 1988.
9. Fey, James T. *Computing and Mathematics: The Impact on Secondary School Curricula.* Reston, VA: National Council of Teachers of Mathematics, 1984.
10. Gardner, Martin. *The Magic Numbers of Dr. Matrix.* Buffalo, NY: Prometheus Books, 1985.
11. Ginsburg, Herbert. *Children's Arithmetic: The Learning Process.* New York, NY: D. Van Nostrand, 1977.
12. Gundlach, Bernard H. "The history of numbers and numerals." In Hallerberg, Arthur E. (Ed.): *Historical Topics for the Mathematics Classroom: Thirty-first Yearbook of the NCTM.* Washington, DC: National Council of Teachers of Mathematics, 1969, 18–36.
13. Hembree, Ray and Dessart, Donald. "Effects of hand-held calculators in precollege mathematics education: A meta-analysis." *Journal for Research in Mathematics Education,* 17 (1986), 83–89.
14. Hiebert, James (Ed.). *Conceptual and Procedural Knowledge: The Case of Mathematics.* Hillsdale, NJ: Lawrence Erlbaum Associates, 1986.
15. Hiebert, James and Behr, Merlyn (Eds.). *Number Concepts and Operations in the Middle Grades.* Reston, VA: National Council of Teachers of Mathematics, 1988.
16. Howden, Hilde. "Teaching number sense." *Arithmetic Teacher,* 36 (1989), 6–11.
17. Kamii, Constance K. *Young Children Reinvent Arithmetic.* New York, NY: Teachers College Press, 1985.
18. Kaput, James. "Quantity structure of algebra word problems: A preliminary analysis." In Lappan, Glenda and Even, Ruhama (Eds.): *Proceedings of the Eighth Annual Meeting of the PME-NA,* 1986, 114–120.
19. Kaput, James. "Representation systems and mathematics." In Janvier, Claude (Ed.): *Problems of Representation in the Learning and Teaching of Mathematics.* Hillsdale, NJ: Lawrence Erlbaum Associates, 1987, 19–26.
20. Kaput, James and Sims-Knight, Judith. "Errors in translations to algebraic equations: Roots and implications." *Focus on Learning Problems in Mathematics,* 5 (1983), 63–78.
21. Kastner, Bernice. "Number sense: The role of measurement applications." *Arithmetic Teacher,* 36 (1989), 40–46.
22. Kenelly, John W. *The Use of Calculators in the Standardized Testing of Mathematics.* Washington, DC: Mathematical Association of America, 1989.
23. Knuth, Donald E. "Algorithmic thinking and mathematical thinking." *American Mathematical Monthly,* 92 (1985), 170–181.
24. Lesh, Richard (Ed.). *Number and Measurement: Papers from a Research Workshop.* Columbus, OH: ERIC/SMEAC, 1975.
25. Maor, Eli. *To Infinity and Beyond: A Cultural History of the Infinite.* Boston, MA: Birkhauser, 1987.
26. Maurer, Stephen B. "Two meanings of algorithmic mathematics." *The Mathematics Teacher,* 77 (1984), 430–435.
27. Myhill, John. "What is a real number?" *American Mathematical Monthly,* 79 (1972), 748–754.
28. Osborne, Alan R. "The mathematical and psychological foundations of measure." In Lesh, Richard (Ed.): *Number and Measurement: Papers from a Research Workshop.* Columbus, OH: ERIC/SMEAC, 1975, 19–45.

29. Pollak, Henry O. (Ed.). "The mathematical sciences curriculum K-12: What is still fundamental and what is not." Washington, DC: Conference Board of the Mathematical Sciences, 1982.
30. Schoenfeld, Alan. *Problem Solving in the Mathematics Curriculum: A Report, Recommendations, and an Annotated Bibliography.* Washington, DC: Mathematical Association of America, 1983.
31. Schwartz, Judah. "Semantic aspects of quantity." Cambridge, MA: Massachusetts Institute of Technology, 1976 (unpublished manuscript).
32. Thomson, Sir William (Lord Kelvin). *Popular Lectures and Addresses.* New York, NY: Macmillan and Co., 1891, 1894.
33. Usiskin, Zal and Bell, Max. *Applying Arithmetic: A Handbook of Applications of Arithmetic.* Chicago, IL: Department of Education, University of Chicago, 1983.
34. Whitney, Hassler. "The mathematics of physical quantities, Part I: Mathematical models for measurement." *American Mathematical Monthly,* 75 (1968), 115–138.
35. Whitney, Hassler. "The mathematics of physical quantities, Part II: Quantity structures and dimensional analysis." *American Mathematical Monthly,* 75 (1968), 227–256.

Uncertainty

~ ~

DAVID S. MOORE

INTRODUCTION

"Uncertainty" is intended to suggest two related topics: *data* and *chance*. Neither is a topic within mathematics; they are both, however, phenomena that are the subject of mathematical study. Roughly speaking, statistics and probability are the mathematical fields that deal with data and chance, respectively.

Recent recommendations concerning school curricula are unanimous in suggesting that statistics and probability should occupy a much more prominent place than has been the case in the past.[12,14] However, because of the emphasis that these recommendations place on data analysis, it is easy to view statistics in particular as a collection of specific skills (or even as a bag of tricks). The task of this essay is not to urge attention to data and chance in the school curriculum—they are already attracting attention—but to develop this strand of mathematical ideas in a way that makes clear the overall themes and strategies within which individual topics find their natural place.

Any discussion that is intended to influence teaching should reflect the experience of teachers and students. Suggestions for curriculum reform detached from that experience offer utopian hopes that are disappointed in practice. Statistics in the schools is not utopian; new material presently being tested is practically useful and aids rather than displaces development of number concepts and skills. Nonetheless, it is easy in our enthusiasm to overlook practical problems and to urge the teaching of subject matter that is unrealistic in quantity or level. It is important to call attention to the difficulties and potential false steps, as well as to the advantages, in using data and chance in the teaching of mathematics. In writing this essay I have tried to err in the practical rather than the utopian direction.

Data

Interest in teaching statistics is certainly due in part to recognition of the place that working with data plays in everyday life and in many occupations. It is increasingly common to teach mathematical topics that are of direct use, rather than to select topics simply because they lead to later topics in mathematics. Statistics is such a topic.

News reports present national economic and social statistics, opinion polls, medical data from both epidemiological studies and clinical trials, and business and financial data. Many citizens must deal with data in more detail on the job. Farmers and agribusiness use crop forecasts and the results of agricultural field trials. Engineers are concerned with data on product performance, quality, and reliability. Manufacturing workers are increasingly asked to record and act on data for process control. The health sciences struggle with data on cost and effectiveness as well as with data from medical research. Business runs on data of every variety: costs, profits, sales projections, market research, and much more. There are compelling practical reasons to learn statistics.

As these examples suggest, data are not merely numbers, but *numbers with a context.* The number 10.3 in the absence of a context carries no information; that the birth weight of a baby is 10.3 pounds enables us to comment on the healthy size of the child. That is, data engage our knowledge of their context so that we can understand and interpret, rather than simply carry out arithmetical operations.

There are, therefore, strong pedagogical as well as practical reasons to teach statistics in the schools. Statistics combines computational activity in a meaningful setting with the exercise of judgment in choosing methods and interpreting results. Statistics in the early grades is taught not primarily for its own sake, but because it is an effective way to develop quantitative understanding and to apply arithmetic and graphing to problem solving.

Teachers who understand that data are numbers in a context will always provide an appropriate context when posing problems for students. Calculating the mean of five numbers is an exercise in arithmetic, not statistics. Calculating the mean price of a popular music tape at five retail outlets is statistics, particularly when combined with a look at the spread in the prices and a comparison with the price of other types of music.

It is essential that the practical and pedagogical advantages of working with data not succumb to an exclusive emphasis on teaching operations. Teachers and developers of curriculum material must exercise imagination in providing data that are meaningful to students. In the upper grades, data from other academic subjects (such as science) can be used, although students rarely connect such data with their everyday life. In the lower grades, data produced by the students themselves are best. Students can produce data in many ways, such as questioning the class ("How many children live in your house?") or by asking each student to measure, count, or estimate some quantity.

The additional effort required to provide data rather than simply numbers should be taken into account when planning instruction. Good data are not just an attractive feature for motivating students; they are essential to the nature of statistics. Yet it is important that the effort required to produce data not overshadow the mathematical ideas taught and learned.

In particular, attempts to produce good data on important issues outside school are always much more difficult than is at first apparent. Unpleasant experiences with time-consuming and confusing attempts to produce data may well discourage teachers from teaching statistics. The difficulties associated with data production activities form the first of several potential barriers to effective reform. Curriculum materials must provide both interesting data and practical, tested suggestions for production of data by students. Over time, teachers can collect and share data sets that pertain to their community and school. Computers are an ideal means of storing and sharing data.

Chance

Some phenomena have predictable outcomes: drop a coin from a known height and the time it takes to fall can be predicted from basic physics. Except for a rather small measurement error, the outcome is certain. If we toss the coin, on the other hand, we cannot predict whether it will show heads or tails. The outcome is uncertain. Yet coin tossing is not haphazard. If we make a large number of tosses, the proportion of heads will be very close to one-half. This long-term regularity is not just a theoretical construct but an observed fact:

- The French naturalist Buffon (1707–1788) tossed a coin 4040 times. Result: 2048 heads, a proportion of $2048/4040 = 0.5069$ of heads.
- Around 1900 the English statistician Karl Pearson heroically tossed a coin 24,000 times. Result: 12,012 heads, a proportion of 0.5005.

- The English mathematician John Kerrich, while imprisoned by the Germans during World War II, tossed a coin 10,000 times. Result: 5067 heads, a proportion of 0.5067.

Phenomena having uncertain individual outcomes but a regular pattern of outcomes in many repetitions are called *random.* "Random" is not a synonym for "haphazard" but a description of a kind of order different from the deterministic one that is popularly associated with science and mathematics. Probability is the branch of mathematics that describes randomness.

The experience of children in and out of school provides less contact with randomness than with data. For example, students do not meet areas of science in which random behavior appears (such as genetics and quantum theory) until secondary school and then only if they elect solid science courses. Uncertainty is of course a pervasive aspect of all human experience; it is the *order* in uncertainty that is hard to observe in casual settings. Even state lotteries, although familiar to many students, give little experience with the orderly aspect of randomness because of their emphasis on extremely unlikely large prizes. These well-publicized games of chance use actual physical randomization but appear to make people rich haphazardly.

Psychologists have shown that our intuition of chance profoundly contradicts the laws of probability that describe actual random behavior. This incorrect understanding is very difficult to correct by formal instruction. Attempts to teach probability and statistical inference without adequate intuitive preparation are a second major pitfall in introducing data and chance into school curricula.

Even at the college level many students fail to understand probability and inference because of misconceptions that are not removed by study of formal rules. The conflict between probability theory and students' view of the world is due at least in part to students' limited contact with randomness. We must therefore prepare the way for the study of chance by providing experience with random behavior early in the mathematics curriculum. Fortunately, the study of data provides a natural setting for such experience. The priority of data analysis over formal probability and inference is an important principle for instruction in uncertainty.

Artificial chance devices (coins, dice, spinners) can be used to produce data in the classroom with the intent of applying data analysis skills to discover the orderly nature of these devices. Uncertainty also appears in data from sources other than chance devices. Repeated measurements of the same quantity (made by several students, for example) yield varying results. Natural variation appears in the heights, reading scores, or incomes of a group of people. It is perhaps surprising that the patterns of variation in careful measurements or in data on many individuals can be described by the same mathematics that describes the outcomes of chance devices.

Experience with variation in data is a first step toward recognizing the connection between statistics and probability. At a later stage the role of deliberate randomization in statistical designs for producing data strengthens this connection. Finally, formal statistical inference uses the language and facts of probability to express the confidence we can have in conclusions drawn from data.

Although the usefulness in everyday life of an understanding of randomness is less obvious than the necessity of dealing with data, practical arguments for teaching about chance are not absent. One goal of instruction about probability is to help students understand that chance variation rather than deterministic causation explains many aspects of the world.

> Suppose that a basketball player over a long season has made 70% of her free throws. At the end of a tournament game she attempts five free throws and makes only two. "Nervousness," say the fans. But this causal explanation need not be correct. A player having a probability of 0.7 of making each shot has a probability of about 0.16 of missing three or more of five shots. Such a performance can easily be simply chance variation.

Some understanding of probability enables us to consider the role of chance rather than seek a specific cause, oftentimes spurious, for every occurrence.

Calculators and Computers

While the advent of fast, easily accessible computing has had an impact on mathematics as a whole, it has revolutionized the practice of statistics. An obvious effect of the revolution is that more complex analyses on larger sets of data are now easy. But the computing revolution has also brought about changes in the nature of statistical practice. In the past statisticians conducted straightforward but computationally tedious analyses based on a specific mathematical model in order to draw conclusions from data. Instruction in statistics showed a corresponding emphasis on learning to carry out lengthy calculations.

Now the paradigm statistical analysis is a dialogue between model and data. The data are allowed to criticize or even falsify the original model. Diagnostic methods to aid this process are a major field of research in statistics. All are computationally intensive, and the most widely adopted make heavy use of graphic display. In addition, freedom from the limits once imposed by hand calculation has led to new methods for inference from even quite small data sets.[3] This changing nature of statistics is readily reflected in instructional styles, especially in increased emphasis on graphical methods and informal data analysis.

The influence of computers has led to some soul searching among mathematicians, some of whom question the nature of a proof based on a computer search of possible cases too numerous for human scrutiny. At a more elementary level, both teachers and parents ask whether early use of calculators will impede understanding of numbers and arithmetic operations. Statisticians, on the other hand, have welcomed calculators and computers as a liberating force. Calculating sums of squares by hand does not increase understanding; it merely numbs the mind. In these circumstances it is natural for a statistician to urge the use of calculators and computers in instruction about data at all levels.

College teaching of statistics already makes universal use of calculators and wide use of statistical software on computers. (There is, of course, a continuum rather than a disjunction between calculators and computers as technology continues its advance.) Here is a typical exercise from basic statistics, reconsidered in the light of easy computing.

> Figure 1 presents a scatterplot of data on the age at which each of a group of children spoke their first word and their later scores on a test of mental ability. Does age at first word help us predict the later test score?

Once upon a time a student would be asked to plot the data and then calculate the least squares regression line (the solid line in Figure 1) together with the correlation coefficient $r = -0.640$. Perhaps the plot would be omitted to save time. Most students would require at least 15 minutes for this exercise with a basic calculator. Only a sadist would ask much more of them.

But it is apparent that the data include two outliers, labeled as cases 18 and 19 in the plot. How do these cases influence the regression analysis? An interactive software package of the kind that is widely available on all varieties of computers provides immediate answers, which can be visually displayed if the computer has graphics capabilities. Case 19, although far from the regression line, does not have a large influence on the position of the line or the value of the correlation r. Case 18, on the other hand, is highly influential. Removing this point moves the regression line to the dashed line in the figure and reduces the correlation to $r = -0.335$, about half its original value. Thus the evidence that age at first word predicts later ability scores is much weaker if case 18 is

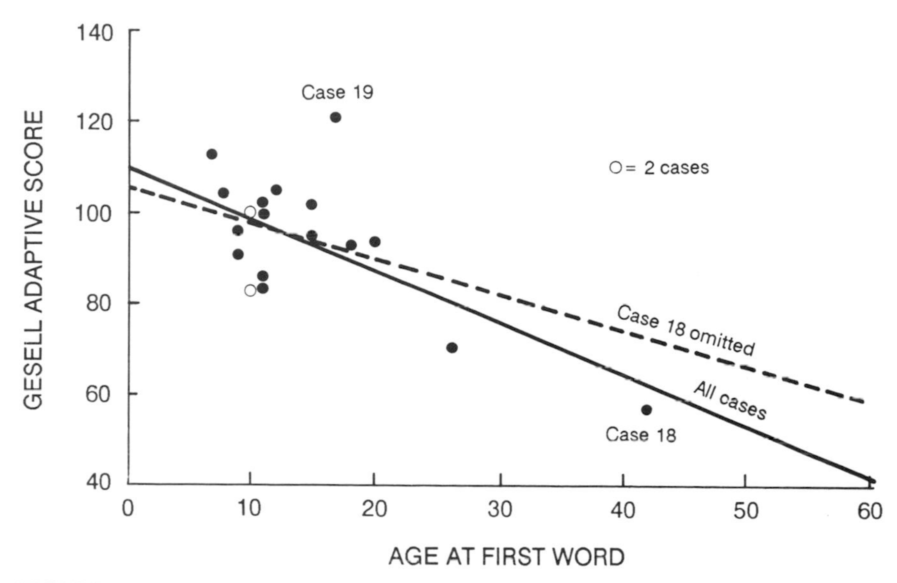

FIGURE 1. Data on the age at which each of 21 children first spoke (horizontal scale) and their Gesell Adaptive Score, the result of an aptitude test taken at a later age. Case 18 is particularly influential in the sense that deleting this point substantially moves the regression line and changes the value of numerical measures such as the correlation.

dropped. (These data are discussed in detail in Examples 3.10 and 3.14 of Moore;[13] most of the figures in this essay are drawn from that text.)

Automating the calculations preserves our energy for a discussion of the data. It is natural for the discussion to take the form of group problem solving: "Is anything unusual? Outlying points. How important are they? Let's try doing the analysis again without them." We are then encouraged to seek additional information about the context of the data—to ask, for example, if the child of case 18 is so slow to begin talking as to be out of place in a study of normal child development. The example also leads us to ask what makes an observation influential, a question that leads to new and important subject matter in statistics.

Automated calculation allows students to concentrate on other aspects of problem solving: planning an appropriate analysis, interpreting the results in their context, and asking new mathematical questions suggested by an exercise. But it is also true that automated calculation can hide the nature of the work that is carried out and impede judgment about whether the work was appropriate to this specific problem. Too often, students believe that computers simply inform us about the truth, as in the *Star Wars* movies.

In a classroom exercise on sampling,[18] students were asked to record the colors of a large sample of M&M candies and to compare the results with computer-produced samples from a uniform distribution of the same colors. The distribution of colors in the candies was far from uniform. The purpose of the exercise was to demonstrate from the comparison that the candy colors were not, in fact, uniformly distributed. Yet "...some students simply believed that the computer model was correct because it was on the computer, even though they had entered the population model themselves."

Overoptimism about the effectiveness of computers is a major potential pitfall in teaching statistics, as is insufficient planning to integrate calculators and computers into the curriculum. Graduated use of calculators and computers is essential if students are to gain their advantages without coming to believe in a "magic box."

Basic arithmetic skills are needed for mental arithmetic and estimation, which are important in checking automated calculations. Four-function calculators preserve control over the order of operations, which must be requested one by one, while automating only the algorithms. A child must understand, for example, the distinction between divisor and dividend in order to use a calculator for long division. A child must know that one finds a mean by adding the observations and dividing by their number in order to compute $\overline{x}$ with a basic calculator. Children can therefore begin to use calculators in their study of data as soon as the operations are understood. Later, a calculator that will compute the sample mean and standard deviation directly from keyed-in data can be used to bypass routine algorithms already mastered.

At a more advanced level, some histograms should be made by hand before turning to attractive software that chooses groups and creates histograms directly from the raw data. Perhaps most importantly, experience with physical chance devices and physical simulations such as drawing colored beads from a box should precede computer simulations. "Microworlds" need have no connection with reality, yet students tend to believe that the computer presents reality. A carefully graduated transition from physical to digital is very important. The practice of graduated use is easiest when calculators and computers are part of the normal classroom environment to be used as needed, not reserved for special projects or upper grades.

From Data to Inference

There are several organizing principles that help us see the mathematical study of data and chance as a coherent whole. One such principle is *the progression of ideas from data analysis to data production to probability to inference.* The discussion in this essay is organized in these same stages:

- Data analysis, which involves organizing, describing, and summarizing data.
- Producing data, usually to answer specific questions about some larger population.
- Probability, the mathematical description of randomness.
- Inference, the drawing of conclusions from data.

This progression of topics represents both the logical development of the field and the level of difficulty of the concepts. It therefore gives the general order in which statistical topics should appear in the school curriculum. Of course, the latter three headings will appear informally from the beginning in the context of data analysis. Experience in producing data—in particular, experience with chance outcomes—can begin in the earliest grades. Similarly, informal conclusions based on data should be encouraged from an early stage.

The main drawback to this outline is that it does not emphasize that probability is important in its own right, not merely as a part of statistics. Both the concept of probability and basic mathematical facts about probability can be introduced in elementary school as soon as fractions are understood. There is, however, a natural place for probability in the progression of statistical ideas. Statistical designs for producing data are characterized by the deliberate use of chance in random sampling and randomized comparative experiments. Here is an opportunity to provide more experience with randomness and to advance to a study of random variation in numerical summaries (such as the mean of several observations). Both physical random selection and simulation can be used.

On the other hand, formal statistical inference requires some understanding of probability. Therefore it makes sense that the section on probability be between those on producing data and inference. Because of the great conceptual difficulties that students encounter in probability and in probability-based inference, formal mathematical treatment of these subjects should probably be an elective rather than a core course in secondary school.

DATA ANALYSIS

Data analysis is descriptive statistics reborn, with new methods, greater emphasis on graphics, and a consistent philosophy due to John Tukey. (Volumes 3 and 4 of Tukey's *Collected Works* contain his writings in this area.[8] A reviewer recommends paper 12 in Volume 4 as a good starting point.) The essence of data analysis is to "let the data speak" by looking for patterns in data without at first considering whether the data are representative of some larger universe.

Inspection of data often uncovers unexpected features. If the data were produced to answer a specific question—this is the setting in which such traditional methods as confidence intervals and significance tests are best justified—the unusual features may lead us to reconsider the analysis we had planned. Careful data analysis therefore precedes formal inference in good statistical practice.

In other cases we do not have specific questions in mind and want to allow the data to suggest conclusions that we can seek to confirm by further study. We then speak of "exploratory data analysis," on the analogy of an explorer entering unknown lands.

The best-known contributions of data analysis are new methods for displaying data, such as stemplots and boxplots (or stem-and-leaf plots and box-and-whisker plots if you prefer longer terms). From these examples it is easy to see data analysis as a collection of clever tools and miss the organizing principles. Both analyses of complex data sets and the order of instruction about data can usefully be guided by three simple principles:

1. Move from simple to complex, from examining a single variable to relations between two variables and connections among many variables.
2. When examining data, look first for an overall pattern and then for marked deviations from that pattern.
3. Move from graphic display to numerical measures of specific aspects of the data to compact mathematical models for the overall pattern.

Displaying Data

The first and third principles suggest that learning about data starts with displaying the distribution of a single variable. Most such data are either counts—that is how qualitative variables such as color become numerical—or measurements with units. Specific methods for data display can advance in parallel with the development of early quantitative concepts. "How many of each color in a bag of M&Ms?" can be determined by counting and displayed with stacks of colored blocks.

Later a stemplot of two-digit numbers can reinforce the distinction between the 10's and the 1's place in whole numbers. A stemplot of two-digit data lists each 10's digit as a "stem" and records the observations by placing their 1's digits as "leaves" on the appropriate stem. Here, for example, is a stemplot of the number of home runs Babe Ruth hit in each of his years with the Yankees.

2	25
3	45
4	1166679
5	449
6	0

Still later we come to histograms. To construct histograms of data with more than a few values requires an understanding of "betweenness" and the ability to group numbers, as well as skill in making and using scales on graphs.

Choice among the available variations on stemplots and histograms requires more judgment as the numbers making up the data become less simple. Stemplots of numbers with several digits often require rounding or truncation, for example. Grouping numbers with several decimal places into classes for a histogram requires a clear understanding of order for decimal numbers. Careful planning is important to avoid inadvertently presenting students with tasks that go beyond their number skills. But it is also clear that data analysis in the elementary grades can reinforce important concepts and skills from the existing mathematics curriculum by applying them in interesting settings.

When we have constructed a display, we must interpret it and communicate our understanding to others. Children are not naturally able to "read" data any more than they are born able to read words. They must be taught both the strategy of looking at data and specific features to be aware of. The strategy is expressed in the second principle: look for pattern, then for deviations. The specific features change as we advance through the stages mentioned in the first principle. An example will illustrate the process in the case of single-variable data.

> In 1961 Yankee outfielder Roger Maris broke Babe Ruth's record of 60 home runs in a single season. Here is a back-to-back comparison of yearly home runs hit by Ruth (on the left) and by Maris during their years with the Yankees:

RUTH		MARIS
	0	8
	1	346
52	2	368
54	3	39
9766611	4	
944	5	
0	6	1

The overall shape of Ruth's distribution is roughly symmetric. The center is at about 46 home runs, in the sense that he hit more than 46 half the time and fewer half the time. There are no strong deviations from the overall pattern. In particular, Ruth's famous 60 home runs in 1927 do not stand out from the other values; it is Babe's best effort but not unusual in the context of his career.

In contrast, Maris's record of 61 homers in 1961 is an *outlier* that falls clearly outside his overall pattern. That overall pattern (excluding the outlier) is again roughly symmetric and is centered at about 23. The different locations of the two distributions show Ruth's general superiority as a home-run hitter.

To see the overall pattern of the distribution of a single variable, we learn to look for symmetry or skewness, for single or multiple peaks, for the center and the degree of spread about the center. Important deviations from a regular pattern include gaps and outliers. Notice that while constructing the display is an operation to be learned, interpretation requires judgment.

No distribution of real data has the perfect mirror symmetry of some mathematical shapes. Not all distributions are well described as either symmetric or skewed. Too much emphasis on classifying what we see will frustrate both teachers and students. Learn to observe marked features, not to debate unclear features. Note also that looking at data naturally leads to attempts to interpret what we see, as when we noticed that Ruth's 60 was not an unusual performance for him, while Maris's 61 was an outstanding achievement far beyond his usual level.

Interpreting the overall shape of a distribution is an important part of learning to look at data. The histogram in Figure 2 displays student-collected data on the lengths of words in *Popular Science* magazine. The

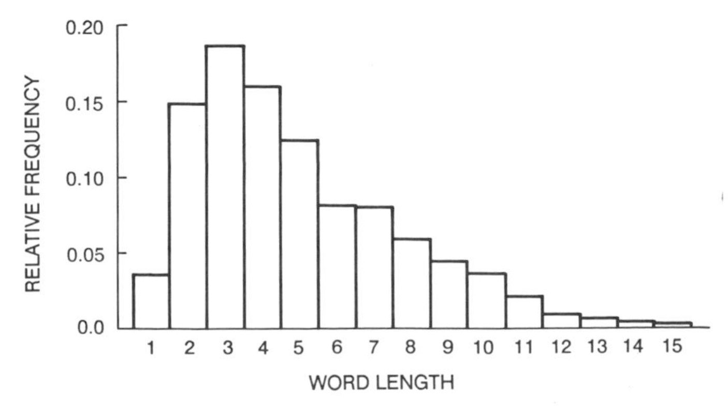

FIGURE 2. Student-collected data on the length of words in *Popular Science* magazine reveal a skewed distribution since shorter words are more common than longer ones.

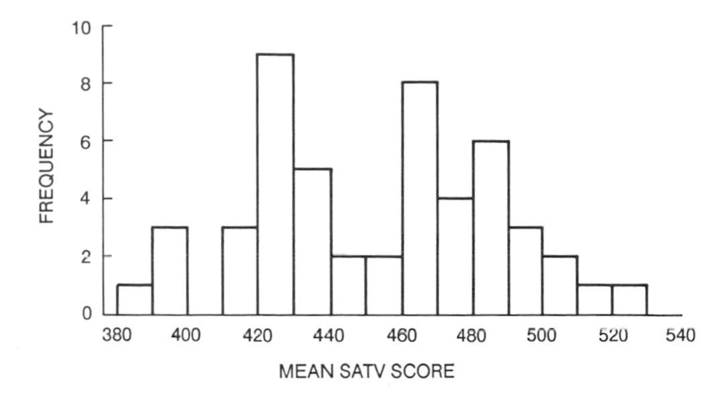

FIGURE 3. Data on the mean verbal SAT score by state reveal a double peak that reflects two different test-taking traditions: in some states most college-bound students take the SAT, whereas in other states only a few do—since the majority take the ACT exam.

distribution is right skewed because there are many two- to five-letter words and fewer long words. (The usual statistical terminology takes the direction of the skewness to be the direction of the longer tail, not the direction in which most observations are concentrated.)

The histogram in Figure 3 shows the mean score by state on the verbal part of the Scholastic Aptitude Test (SAT). This distribution is double peaked. The peak near 425 represents states in which most college-bound students take the SAT; the higher-valued peak represents states in which most students take the American College Testing (ACT) examination and only students applying to selective colleges take the SAT.

Numerical Description

Already in examining the Ruth and Maris home-run data we saw that calculation can help us describe data. By simple counting ("half more and half less") we can give numbers that make more exact the difference in centers that we see in the stemplots. The natural progression of mathematical tools is expressed in the third organizing principle: graphics to numerical measures to mathematical models.

In the case of the distribution of values of a single variable, the basic aspects to be described numerically are the *center* (or *location*) and the *spread* (or *dispersion*) of the distribution. (The older term "central tendency," which is both longer and less clear than "center" or "location," is rarely used by statisticians and should be abandoned.) There are two common sets of descriptive measures for location and spread: the median with the quartiles (or perhaps other percentiles) and the mean with the standard deviation. Percentiles require only counting and an understanding of simple fractions (1/4, 1/2, 3/4 for median and quartiles). The mean is the arithmetic average. So the mean, median, quartiles, and smallest and largest values can be introduced as students develop basic arithmetic skills. These simple measures form a helpful descriptive vocabulary.

Experience with the connection between the shape of displayed data and numerical measures strengthens number sense. Although both the displays and the measures seem elementary, the amount of mathematical understanding required to use them effectively (as opposed to simply calculating the measures) should not be underestimated. In one field test of new teaching material, for example, neither students nor the teacher could believe that adding observations to the right end of a particular distribution with many tied observations in the center left the median unchanged.[19] Hands-on experience with many sets of data, including attempts to estimate measures by looking at the display and discussing results, helps students construct their own understanding of such apparently simple operations as counting halfway up the ordered list (the median) and averaging all the values (the mean).

Numerical description of a distribution by the median, quartiles, and extreme observations leads to a new graphic display, the *boxplot.* An example shows how useful this device can be. U.S. Department of Agriculture regulations group hot dogs into three types: beef, meat, and poultry. Do these types differ in the number of calories they contain? In Figure 4 three boxplots display the distribution of calories per hot dog

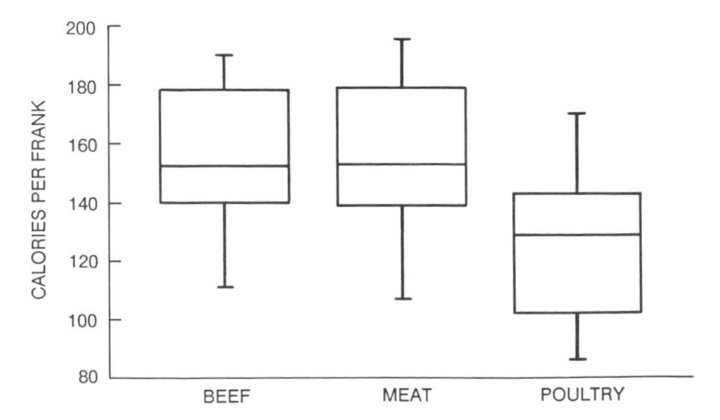

FIGURE 4. Three boxplots display visually the median, quartiles, and extremes of calories provided by various brands of hot dogs belonging to three standard types: beef, meat, poultry. One can easily see that poultry hot dogs as a group contain fewer calories per hot dog.

among brands of the three types. The box ends mark the quartiles, the line within the box is the median, and the whiskers extend to the smallest and largest individual observations. We see that beef and meat hot dogs are similar but that poultry hot dogs as a group show considerably fewer calories per hot dog.

Mathematical Models

In this brief discussion of single-variable data, we have not yet mentioned either the standard deviation or the final stage in the progression from graphical display to numerical description to mathematical model. The standard deviation has several disadvantages for data description. It is unpleasant to calculate with a basic calculator, is very sensitive to a few extreme values, and is difficult to motivate clearly. (The mean—or median—of the absolute deviations of the observations from their mean is preferable on all three counts.)

Yet the standard deviation is very important in statistics, mainly because it is the natural measure of spread for normal distributions. Normal curves provide an example of a compact mathematical description of the overall pattern of a distribution of data. They are mathematical idealizations that do not catch the irregularity of real data or deviations such as outliers. Normal curves are, for example, perfectly symmetric.

Most curriculum materials intended for general students stop short of presenting normal distributions. This is true, for example, of the Quantitative Literacy series[7,10,11,15] developed jointly by the American Statistical Association and the National Council of Teachers of Mathematics. One reason may be the traditional view of normal and other distributions as *probability* distributions, to be developed only after considerable study of probability. But it is not necessary to introduce formal probability to suggest that the heights of a large group of people of similar age and sex are roughly normal or that the stopping point of a spinner is roughly uniform over a circle.

Figure 5 shows a histogram of the Iowa Test vocabulary scores of all 947 seventh-grade students in Gary, Indiana, with the normal curve that approximately describes the distribution of scores. It shows quite clearly how a normal curve provides an idealized mathematical model for certain distributions of data.

Moving from particular observations to an idealized description of "all observations" is a substantial abstraction. The use of a mathematical model such as a normal or uniform distribution to formulate this abstraction is a substantial step toward understanding the power of mathematics. Computer simulation is quite helpful at this point.

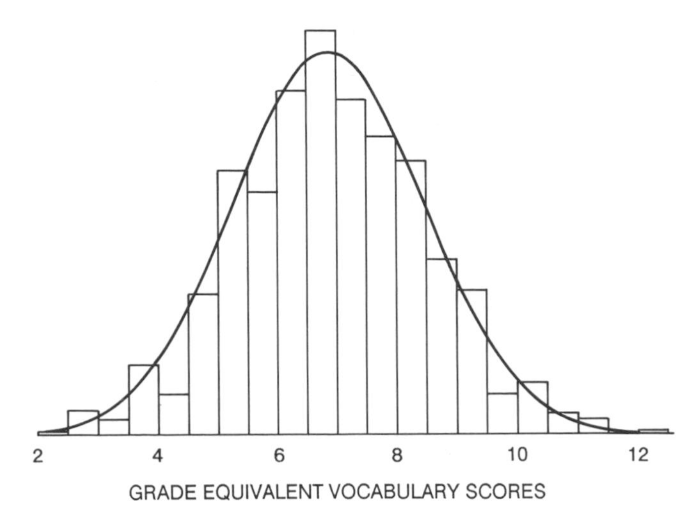

FIGURE 5. A histogram of vocabulary scores of nearly 1000 seventh-grade students shows close adherence to the idealized distribution of the bell-shaped normal curve.

Students can formulate a "population model" on the basis of their experience with data, enter their model into the computer, and simulate observations from the population. Comparing simulated data to the model provides more experience with probability and randomness. The basic properties of normal curves, the idea of standardizing observations to the scale of standard deviation units about the mean, and the use of the standard normal table to calculate relative frequencies can be developed in the setting of models for regular patterns in data.

Although distributions in the mathematical sense complete the progression of descriptive methods for single-variable data, they must appear rather late even when it is understood that distributions can appear before a full introduction to probability. Meanwhile, experience with several-variable data would have been advancing as students develop the necessary mathematical concepts and skills. The beginning study of two-variable data comes later than examination of a single variable, in accordance with our first principle, but usable mathematical models are more accessible in the two-variable case.

The basic graph for two-variable data is the scatterplot, which provides a setting for understanding coordinates in the plane. Clusters (female and male students?) and outliers in a scatterplot provoke discussion. The simplest overall pattern is a linear trend. The mathematical model that gives a simple description of a linear pattern is a straight line with its equation. Numerical measures include measures of the center and spread of each variable separately, the slope of a fitted line as a description of linear relationship, and perhaps the correlation coefficient as a measure of the strength of linear association.

The correlation coefficient, like the standard deviation, is tied to traditional statistical models and methods whose advantages, while real, are not clear until a quite advanced stage of study. The correlation coefficient is closely related to least squares regression; that is, correlation measures the strength of a specific kind of straight-line association. Just as the standard deviation should be delayed until normal distributions give it a context, correlation and least squares regression need not make their appearance until secondary school students undertake a substantial study of statistics for its own sake.

Much of data analysis, while useful in its own right, can be taught from early elementary school through the first years of secondary school as part of the general effort to develop quantitative skills and reasoning. In this setting, straight lines can be fit by eye or by simple methods that are computationally easier than least squares and more resistant to extreme observations. The Quantitative Literacy material[10] offers a clear explanation of such methods for use in the middle grades.

Other aspects of several variable data deserve priority over correlation and least squares regression. These include the distinction between explanatory and response variables, the relation of association to causation, and the effects of unmeasured "lurking variables" on an observed association. These ideas are subtle but not computational; they are best grasped by guided experience with and discussion about actual data, using a variety of display and computational methods; and they are closely related to an understanding of the kinds of explanations offered by the natural and social sciences.

In teaching data analysis in a general school curriculum, topics should be chosen not for their importance in the discipline of statistics but for their immediate relevance to students, their usefulness in strengthening general quantitative understanding, and their contribution to developing reasoning about uncertain data. Statistics *is* important in its own right—more important than calculus in most occupations—and that importance should be reflected in a substantial elective course in the upper secondary years that includes more advanced data analysis as well as data production, probability, and inference.

PRODUCING DATA

Good data are as much a product of intelligent human effort as are compact disc players and hybrid corn. There are several reasons why producing data is an important part of teaching about data and chance.

Data analysis is most effectively carried out on data with which we are intimately familiar, for familiarity suggests both expected features to look for and explanations for unexpected features. Statistical designs for producing data to answer specific questions are the conceptual bridge linking data analysis to classical probability-based inference. And there is no better cure for the extreme attitudes—either unwarranted cynicism or misplaced trust—with which statistical evidence is often greeted than experience that begins with a question and ends with answers based on data that we ourselves have produced.

Data used in the teaching of statistics come from several sources. Much of it is *provided data,* numbers simply provided by the teacher or the text. With concerted effort to choose data on topics within students' experience or interests and to provide appropriate background information, provided data can offer a good setting for interpretation and discussion as well as for building skills. Provided data are more useful with older children who have the wider knowledge and experience to understand the context of the data. Interesting information that students could not produce themselves can be put before the class, and the time and effort saved can be well used. Government data on nearby towns or neighborhoods, for example, often show patterns in population, housing, income, and health that are informative and surprising.

A second category, *class data,* is collected in the classroom and is relevant primarily to students in the class without raising the question of whether conclusions about some larger population are warranted. Class data provide a natural setting for teaching data analysis, which has a similar restriction on the scope of its conclusions. Simple questions are a beginning: "How many children live in your house?" "How much money do you have in your pocket?" The first question produces whole number data, the second two-place decimals. Planning the production of data involves thinking ahead to the analysis that will be called for, a reminder as relevant to professionals armed with software as to teachers attentive to whether their students should face counts or decimals. Measurements can also produce class data: with a tape measure, find the shoulder width and armspan of all the students, then make a scatterplot and study the relationship revealed.

Experiments are a third source of data. Experimentation is active data production. Observation, whether questioning or measuring, seeks to collect data without changing the people or things observed. In an experiment we actually apply some stimulus in order to observe the response. The distinction between explanatory and response variables—an essential part of causal explanations—is clearest in the setting of an experiment. The experiments most familiar in basic science, unlike the questions or measurements that produce class data, do invite conclusions that apply to the world at large. When students heat a closed volume of air and watch a balloon expand, they are asked to understand not just the behavior of the one balloon but also the effect of heat on gases in general. This rather large conceptual leap is often left implicit.

Moving from class data to statistically designed *samples* has the great advantage of making explicit the transition from data about this one class to data that represent a larger population. How to sample is a topic within statistics, with implications far broader than merely generating attractive data for analysis. Statistics also has much to say about how to experiment, although the advice is not relevant to most experiments in basic science. The design of samples and of experiments is a major topic in the systematic study of data production. But another topic comes first, both logically and in classroom experience: asking questions and measuring to produce class data both raise the issue of measurement.

Measurement

To measure a characteristic means to represent it by a number. This basic notion already introduces an abstraction. Thinking about measurement leads at length to a mature grasp of why some numbers are informative and others are irrelevant or nonsensical. First, what is a *valid* (appropriate or meaningful) way to measure a particular characteristic? Begin with tangible physical characteristics. Length is easy—we agree that a ruler will do it. Area is harder, because we have no device that we can "put beside" the many shapes possible in two dimensions as we put a ruler beside any length. We must concern ourselves with understanding the characteristic to be measured, with devising a satisfactory instrument, and with the units that result and their relations to other units. Even for physical measurements the study of these questions extends throughout the school years both in mathematics and in science.

But the validity of physical measurements is simple compared with the measurement problems of the social and behavioral sciences. What is a good way to measure how rich a family is or the friendliness of a fellow student? What do the Iowa Tests or the ACT and SAT college entrance examinations really measure? A detailed examination of such questions would lead too far afield. But students should be encouraged always to ask whether data are in fact valid for the proposed use. Drivers over 65 years of age are involved in more fatal accidents than drivers aged 16 and 17. So teens aren't so risky after all? No—there are many more drivers over 65. The *rate* rather than the *count* of accidents is the appropriate measurement, and the fatal accident rate for teens is about three times that for the elderly.

The second major aspect of the quality of measurements, after validity, is *accuracy.* A measuring process may show systematic error, or *bias,* as when a scale always reads 3 pounds low. Bias is a straightforward idea only when the "true value" that measurement should yield is clearly understood. Possible bias in SAT scores is a continuing source of intense debate, since no "correct" value is available for comparison. As usual, physical measurement is much more straightforward than behavioral or social measurement.

A measuring process also shows *variation;* that is, repeated measurements of the same quantity do not give identical results. The variations in common instruments such as bathroom scales and tape measures are small relative to the desired accuracy, so we are accustomed to ignoring variation in measurement. Activities that demonstrate measurement variation are needed. Requiring students to interpolate between scale markings when measuring length or weight, or to estimate a length or count by eye, provides a set of varying measurements whose distribution can be displayed and discussed with the tools of data analysis. Bias is described by the center of the distribution of measurements and variation by the spread.

Measurement activities followed by discussion of the data they produce increase students' sensitivity to the issue of the quality of measurements. Here is an example from a college class.

> The instructor asked each student to measure and record his or her pulse rate (heartbeats per minute) on a piece of paper. A stemplot of the collected data showed an outlier that almost certainly resulted from a gross error, though no one would admit having recorded a seated pulse rate of 180. The stemplot also showed a suspicious concentration of pulse rates ending in 0. Questioning revealed that several students had learned in aerobics classes to count beats for 6 seconds and multiply by 10. This led to a discussion of the measurement methods used. Most students had counted beats for 60 seconds. The class decided that this is more accurate than the aerobics class method, but it suffers from partial beats at the beginning and end of the 60-second period. Someone suggested timing exactly 50 beats with a stopwatch and calculating beats per minute from this time. This was accepted as a more accurate practical measurement method.

Statistical Designs

Design of sample surveys and experiments is a core topic in statistics and a major transition in concepts. Data analysis emphasizes understanding the specific data at hand. Now the data are regarded as representing a larger population. It is the population we seek to understand. Students do not find this added abstraction easy to assimilate. They persist, for example, in trying to explain variable results when an experimental task is carried out by several students in terms of individual characteristics of Sarah, Matthew, and Ruth. The "sampling" point of view regards these students as representative of a large population of students. We are no longer interested in individual features that may explain the performance of Sarah, Matthew, and Ruth.

The transition from data analysis to inference follows a parallel path in mathematical abstraction. The sample mean $\overline{x}$ is no longer just a single number, a measure of location for these data. It is a realization of a random variable to be considered against the background of the distribution of the random variable; it must be viewed against what would happen if we repeated the data production process many times. The difficulty of these new ideas cannot be disguised.

Fortunately, the intimate connection of designed data production with the ideas of probability and the logic of inference need not appear at once. There is much valuable insight into data to be gained first. It is very important, for example, to recognize unrepresentative data. Anecdotal evidence based on a few individual cases known to us influences our thinking in ways that cannot withstand examination and therefore must be examined. Individual cases catch our attention because they are unusual in some way or because they occur in our immediate environment. Examples and discussion will show that there is no reason to expect these cases to be in any way typical.

Improper sampling methods, especially voluntary response samples in which the respondents choose themselves, are also fair game. Here is an example:

> Advice columnist Ann Landers conducts a voluntary response survey every few years by asking her readers to respond to a provocative question. The results are always good for news articles and radio interviews that publicize her column. Her first survey is the most instructive because a comparison is available. In 1975 Ann Landers asked "If you had it to do over again, would you have children?" Almost 70% of the nearly 10,000 respondents said "no." Many accompanied their responses by heart-rending tales of the cruelties inflicted on them by their

> children. It is the nature of voluntary response to attract people with strong feelings, especially negative feelings, about the issue in question. A nationwide random sample commissioned in reaction to the attention paid to Ann Landers's results found that 91% of parents *would* have children again.

Voluntary response can easily produce 70% "no" when the truth is 90% "yes." Such data carry no useful information about anyone except the people who stepped forward. Yet the news media not only report voluntary response data as if they described a general population, they also operate call-in and write-in polls that produce more such data. Alert students will easily find examples. Discussion of anecdotal evidence and voluntary response makes clear the need for a systematic method for selecting samples.

The statistician's recommended method is to let impersonal chance select the sample. Random sampling eliminates the biases of personal choice, whether by the sampler or by the respondents. The deliberate use of chance is the most important statistical principle for producing data. It seems at first unnatural to abandon human judgment, but chance appears less outrageous when set against anecdotal evidence and voluntary response. The use of chance is illustrated by simple random samples, which give all possible samples of the stated size the same chance to be the sample actually chosen.

Simple random samples are easy to experience in the classroom, first by drawing names from a hat or varicolored beads from a sampling bowl. Use of a random number table follows, and finally computer simulation. Do recall the warning that too rapid introduction of the computer will obscure the nature of random selection. The more elaborate random sampling designs used in national sample surveys need not appear in introductory instruction.

The simplest randomized comparative experiments are closely related to simple random samples. Once again the need for good design can be made apparent by discussion of some uncontrolled or unrandomized experiments. Here is an example:

> A political scientist interested in the effectiveness of propaganda in changing opinions conducted an experiment with student subjects. The students took a test of their attitude toward Germany, then read German propaganda regularly for several months, after which their attitude was again measured. The year was 1940. Between test and retest, Germany invaded and conquered Holland and France. The students' attitude toward Germany changed drastically, but we shall never know how much of this change was due to reading German propaganda.

The design of this experiment had a form familiar in laboratory experiments in the natural sciences:

Observation ⟶ Treatment ⟶ Observation

Outside the controlled environment of the laboratory, experiments with such simple designs often fail to yield useful data. The effect of the treatment cannot be distinguished from the effect of external variables, though not all such disturbances are as dramatic as the fall of France.

Statistically designed experiments involve two basic principles: comparison (or *control*) and randomization. The simplest randomized comparative design compares two treatments, one of which may simply be a control treatment such as not reading propaganda. Here is the design in outline:

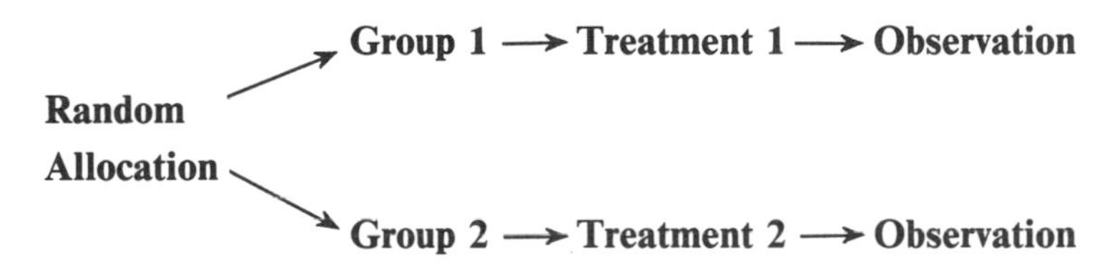

The random allocation assigns a simple random sample of the subjects to Treatment 1; the remaining subjects receive Treatment 2. Randomization assures that there is no bias in assigning subjects to treatments. The groups are therefore similar (on the average) before the treatments are imposed. Comparison assures that outside forces act equally on both groups. If care is taken to treat all subjects similarly except for the experimental treatments, any systematic difference in response must reflect the effect of the treatments. The logic of comparative randomized experiments allows conclusions about causation—the response is not just associated with the treatment but is actually caused by it.

As in the case of sampling, more elaborate designs are common in practice but need not appear in beginning instruction. Classroom experience with randomization is easy and valuable. Consider, for example, tokens such as gumdrop figures that represent subjects to be assigned to two competing treatments for severe headaches. Students carry out the random assignment. Some of the tokens bear a mark on the bottom, invisible when the randomization is done. These subjects, unknown to the experimenters, have a brain tumor that will render any treatment ineffective. How evenly did randomization divide these subjects between the two groups? Do the randomization repeatedly and display the distribution of counts. Repeated randomization provides experience with random variation that leads toward probability and inference.

Some Cautions

With the fundamentals of both data analysis and data production in hand, older students can contemplate serious statistical studies. Examples from recent curriculum projects include a sample of student opinion about the selections served in the school cafeteria; a sample of vehicles at a local intersection, classified by type and home county as revealed by the license plate; and an experiment on the effect of distance and angle on success in shooting a nerf basketball. The design of such studies provides valuable experience in applying statistical ideas. Analysis of real data to arrive at solid conclusions is satisfying. But the practical problems of producing the data must be anticipated and kept within acceptable limits.

Here is an excerpt from a report of a careful study[1] of new statistics material for secondary schools. Some of the data production activities were quite elaborate, including both the road traffic survey and the nerf basketball experiment. Their experience is cautionary.

> Our field test experiences have convinced us that data collection is an important component of statistics education for at least two reasons. First, learning how to design and conduct data collection activities (e.g., determining independent and dependent variables and sample size) is fundamental to statistics. Second, data collection is a motivating experience that makes statistical analysis more meaningful and interesting to students.
>
> Our experiences also convinced us, however, that data collection can present some formidable challenges in the classroom. For example, our field test teachers report that they spent an inordinate amount of class time collecting data as opposed to exploring and analyzing data, only to find that students' data was incomplete or inaccurate. These challenges proved to be so disruptive to academic progress that the teachers grew reluctant to conduct statistical investigations that depended on data collection.

PROBABILITY

Chance variation can be investigated empirically, applying the tools of data analysis to display the regularity in random outcomes. Probability gives a body of mathematics that describes chance in much more detail than observation can hope to discover. Probability theory is an impressive demonstration of the power of mathematics to deduce extensive and unexpected results from simple assumptions.

Coin tossing, for example, is described simply as a sequence of independent trials each yielding a head with probability 1/2. From this unassuming foundation follow such beautiful results as the law of the iterated logarithm, which gives a precise boundary for the fluctuations in the count of heads as tossing continues. The distribution of the count

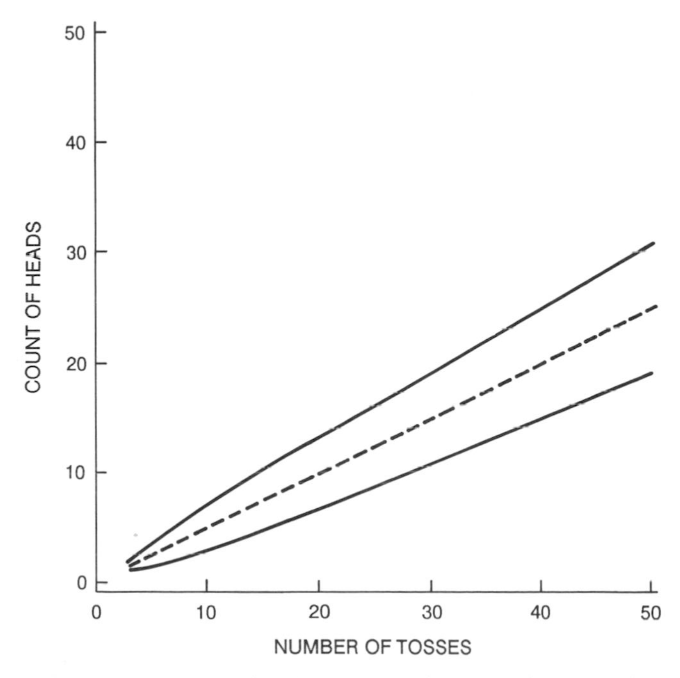

FIGURE 6. The law of the iterated logarithm describes the region of fluctuations in coin tossing: the center line is the mean $n/2$, bounded on either side by curves whose distance from the center line is $\frac{1}{2}\sqrt{n}\sqrt{2\log\log n}$.

of heads after n tosses of a fair coin has a mean of $n/2$, which when plotted against n appears as a straight line (see Figure 6). The standard deviation of the count of heads in n tosses is $0.5\sqrt{n}$. The law of the iterated logarithm says that fluctuations in the count of heads extend $\sqrt{2\log\log n}$ standard deviations on either side of the mean. The count of heads plotted against n will approach within any given distance of this boundary infinitely often as tossing continues, but will cross it only finitely often. Data analysis, even aided by computer simulation, could never discover the law of the iterated logarithm.

As with other beautiful and useful areas of mathematics, probability has in practice only a limited place in even secondary school instruction. Because the fundamentals of probability are mathematically rather simple, it is easy to overlook the extent to which the concepts of probability conflict with intuitive ideas that are firmly set and difficult to dislodge by the time students reach secondary school. Misconceptions often persist even when students can answer typical test questions correctly. The conceptual difficulty of probability ideas is affirmed by both the experience of teachers and by research.[5,21]

Guided experience with randomness in earlier years is an important prerequisite to successful teaching of formal probability. It is no accident that mathematical probability originated in the study of games of chance, one of the few settings in which simple random phenomena are observed often enough to display clear long-term patterns. Teaching can attempt to recapitulate this historical development by recording data from chance devices and later from random sampling and computer simulations. But no matter whether such experience occurs early or late in a student's development, it takes significant time to gain appropriate insight into the behavior of random events.

Basics

The first steps toward mathematical probability take place in the context of data from chance devices in the early grades. Learn to look at the overall pattern and not attempt a causal explanation of each outcome ("She didn't push the spinner very hard"). This abstraction is made easier because looking for the overall pattern of data is one of the core strategies of data analysis.

Next recognize that, although counts of outcomes increase with added trials, the proportions (or relative frequencies) of trials on which each outcome occurs stabilize in the long run. Probabilities are the mathematical idealization of these stable long-term relative frequencies. As students learn the mathematics of proportions, study of probability can begin with assignments of probabilities to finite sets of outcomes and comparison of observed proportions to these probabilities.

Comparison of outcomes to probabilities can be frustrating if not carefully planned. Computer simulation is very helpful in providing the large number of trials required if observed relative frequencies are to be reliably close to probabilities. In short sequences of trials, the deviations of observed results from probabilities will often seem large to students. Psychologists[20] have noted our tendency to believe that the regularity described by probability applies even to short sequences of random outcomes. This belief in an incorrect "law of small numbers" explains the behavior of gamblers who see a run of winning throws with dice as evidence that the player is "hot," a causal explanation offered because we greatly underestimate the probability of runs in random sequences.

> Ask several people to write down a sequence of heads and tails that imitates 10 tosses of a balanced coin. How long was the longest run of consecutive heads or consecutive tails? Most people will write a sequence with no runs of more than two consecutive heads or tails. But in fact the probability of a run

of three or more heads in 10 independent tosses of a fair coin is 0.508, and the probability of either a run of at least three heads or a run of at least three tails is greater than 0.8.

Probability calculations involving runs are quite difficult—this is a good area for computer simulation. The runs of consecutive heads or consecutive tails that appear in real coin tossing (and are predicted by probability theory) seem surprising to us. Since we don't expect to see long runs, we may conclude that the coin tosses are not independent or that some influence is disturbing the random behavior of the coin.

The same misconception appears on the basketball court. If a player makes several consecutive shots, both fans and teammates believe that he or she has a "hot hand" and is more likely to make the next shot. Yet examination of shooting data[22] shows that runs of baskets made or missed are no more frequent than would be expected in a sequence of independent random trials. Shooting a basketball is like throwing dice, though of course the probability of making a shot varies from player to player. As these examples suggest, even the idea of probability as long-term relative frequency is quite sophisticated and needs careful empirical backing.

Somewhat later a thorough understanding of proportions motivates the mathematical model for probability: a sample space (set of all possible outcomes) and an assignment of probability satisfying a few basic laws or axioms that include the addition rule $P(A \text{ or } B) = P(A) + P(B)$ for disjoint events. Further additive laws for simple combinations of events can be derived from these or, more simply, motivated directly from the behavior of proportions. These additive laws are the mathematical content of elementary probability.

At this point in the development of mathematical probability, let us pause for some nonnumerical exercises that apply probability laws along with another aspect of mathematical thinking that is not natural in students: careful and literal reading of logical statements. Psychologists studying probability concepts offer many exercises that reveal misconceptions and can help to correct them. For example, Tversky and Kahneman[21] presented college students with a personality sketch of a young woman and then asked which of these statements was more probable:

- Linda is a bank teller.
- Linda is a bank teller and is active in the feminist movement.

About 85% of the students chose the second statement, even though this event is a subset of the first. This error persisted despite various attempts at alternative presentations that might make the issue more transparent. The subjects had not studied probability. "Only" 36% of social science graduate students with several statistics courses to their credit gave the wrong answer in a similar trial. There is thus some hope that study helps us recognize the relevance of mathematical facts about probability in everyday thinking. Nisbett et al.[17] report before-and-after comparisons that provide stronger evidence of the effect of formal study. Emphasis on the conceptual and qualitative aspects of probabilistic thinking, both prior to and in company with study of the mathematics of probability, is most worthwhile.

Further Study

The development of substantial applicable skills, as opposed to a basic conceptual grasp of probability, requires more detailed study. At this point we leave the core domain of mathematical concepts to which all students should be exposed. There are several logical paths into intermediate probability. The choice of material will depend, for example, on whether probability will be pursued as an important topic in its own right or whether it is intended primarily to lead to statistical inference.

First, a negative recommendation: do not dwell on combinatorial methods for calculating probabilities in finite sample spaces. Combinatorics is a different—and harder—subject than probability. Students at all levels find combinatorial problems confusing and difficult. The study of combinatorics does not advance a conceptual understanding of chance and yields less return than other topics in developing the ability to use probability modeling. In most cases all but the simplest counting problems should be avoided.

A more fruitful step forward from the basics of probability is to consider conditional probability, independence, and multiplication rules. Knowledge of the occurrence of an event A often modifies the probability assigned to another event B. For example, knowing that a randomly selected university professor is female reduces the probability that the professor's field is mathematics. The conditional probability of B given A, denoted by $P(B|A)$, need not be equal to $P(B)$; if the two are equal, events A and B are independent. These notions involve both new ideas and basic skills that are invaluable in constructing probability models in the natural and social sciences.

It is quite possible to present the idea of independence and the multiplication rule $P(A \text{ and } B) = P(A)P(B)$ for independent events with little if any attention to conditional probability in general. This path is attractive if the goal is to reach statistical inference most efficiently and also avoids the considerable conceptual difficulties associated with conditional probability. The binomial distributions for the count of successes in a fixed number of independent trials are quickly within reach, as are other interesting applications such as reliability of complex systems.

If conditional probability is avoided, stress the qualitative meaning of independence and the danger of casually assuming that independence holds. The essay by Kruskal[9] contains examples and reflections on the casual assumption of independence, with emphasis on "independent" testimony to alleged miracles. Topics related to independence, to binomial distributions, and to the multiplication rule for independent events should be staples of upper-grade secondary mathematics.

A careful study of conditional probability is attractive when the goal is to enable students to construct and use mathematical descriptions of processes at a moderately advanced level. Modeling of multistage processes that are not deterministic requires conditional probabilities. To give only a single example, the issue of false positives in testing for rare conditions applies conditional probability to questions as current as testing for drugs, the use of lie detectors, and screening for AIDS antibodies. Here is an example based on a recent report,[6] where a detailed statistical analysis can be found:

> The ELISA test was introduced in the mid-1980s to screen donated blood for the presence of AIDS antibodies. When antibodies are present, ELISA is positive with a probability of about 0.98; when the blood tested is not contaminated with antibodies, the test gives a positive result with a probability of about 0.07. These numbers are conditional probabilities. If one in a thousand of the units of blood screened by ELISA contain AIDS antibodies, then 98.6% of all positive responses will be false positives.

The calculation of the prevalence of false positives among ELISA blood screening tests for AIDS antibodies can be carried out with a simple tree diagram such as that displayed in Figure 7. Students armed with an understanding of conditional probability and tree diagrams can easily program computer simulation of processes too complex to study analytically.

Conditional probability brings a new set of conceptual difficulties that, like those in the early study of probability, can be easily and unwisely overlooked if instruction is overly directed at teaching definitions and rules. Students find the distinctions among $P(A|B)$, $P(B|A)$, and $P(A \text{ and } B)$ hard to grasp. Display a photograph of an attractive and well-dressed woman and ask the probability that she is a fashion model. The answers show that the question is interpreted as asking the conditional probability that a woman known to be a fashion model is attractive and well dressed. That is, respondents confuse $P(A|B)$ and $P(B|A)$.

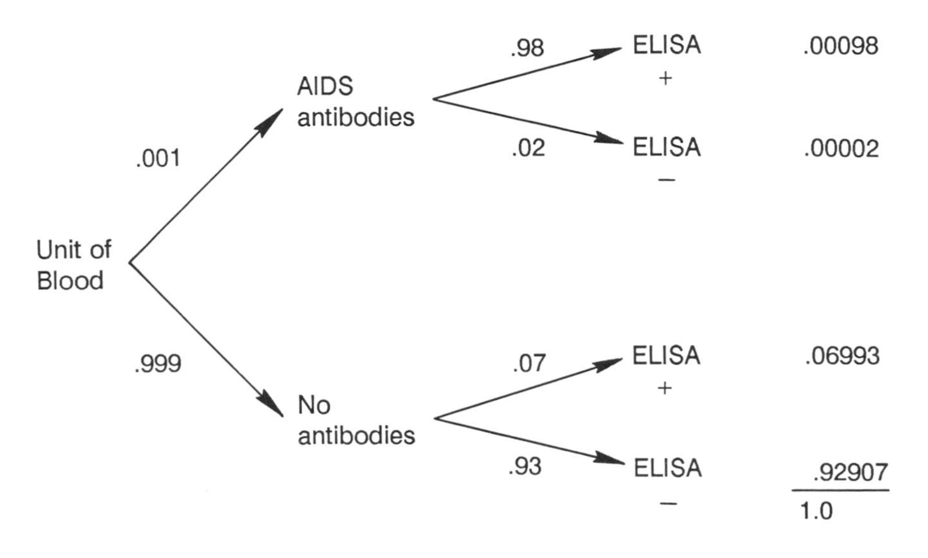

$$P(\text{NO AIDS} \mid \text{ELISA} +) = \frac{P(\text{NO AIDS \& ELISA} +)}{P(\text{ELISA} +)}$$

$$= \frac{.06993}{.00098 + .06993}$$

$$= \frac{.06993}{.07091} = .986$$

FIGURE 7. The calculation of false positives in the ELISA test for AIDS antibodies can be carried out in a tree diagram in which the appropriate conditional probabilities are multiplied along each branch.

Qualitative exercises in identifying the information A that is known and the event B whose probability is wanted are an essential preliminary to formal work with $P(B|A)$.

Transition to Inference

Random sampling and experimental randomization provide experience with randomness that motivates not only the study of probability but also the reasoning of probability-based inference. Repeated sampling or repeated experimental randomization clearly produces variable results. This variation is random in the technical sense, rather than haphazard, because the design uses an explicit chance mechanism. So an opinion poll's conclusion that 61% of all American adults want a national health insurance system requires a margin of error that reflects the probable degree of random variation in similar sample surveys. Similarly, the conclusion that a new medical treatment outperforms a standard treatment can be sustained only if the margin of superiority exceeds the probable random variation in similar experiments.

The random outcomes observed from data production are statistics such as sample proportions and sample means. Sample statistics are *random variables* (random phenomena having numerical values). The regular long-term behavior of such statistics in repeated sampling or repeated experimental randomization is described by a sampling distribution. It is usual to view sampling distributions as probability distributions of random variables. Random variables, their distributions, and their moments make up another block of material in intermediate probability.

Proportions involve the distribution of a count, which is binomial under slightly idealized assumptions. Sample means have a normal distribution if the population distribution is normal. General rules for manipulating means and variances of random variables apply to sample proportions and means. In particular, the standard deviations of sample proportions and means both decrease at the rate $1/\sqrt{n}$ as the sample size n increases, a fact that leads to an understanding of the advantages of larger samples.

What happens as the number n of observations grows without bound? The major limit theorems of probability address this question. The law of large numbers says that sample proportions and means approach (in various senses) the corresponding proportions and means in the underlying population. The central limit theorem says that both proportions and means become approximately normally distributed as the sample size grows.

Figure 8 illustrates the central limit theorem in graphical form. It begins with the distribution of a single observation that is right skewed and far from normal. Distributions of this form are often used to describe the lifetime in service of parts that do not wear out. The mean of this particular distribution is 1. The other curves in the figure show the distribution of the mean $\overline{x}$ of samples of size 2 and of size 10 drawn from the original distribution. The characteristic normal shape is already starting to emerge when only 10 observations are averaged. A computer simulation could show the effect even more dramatically.

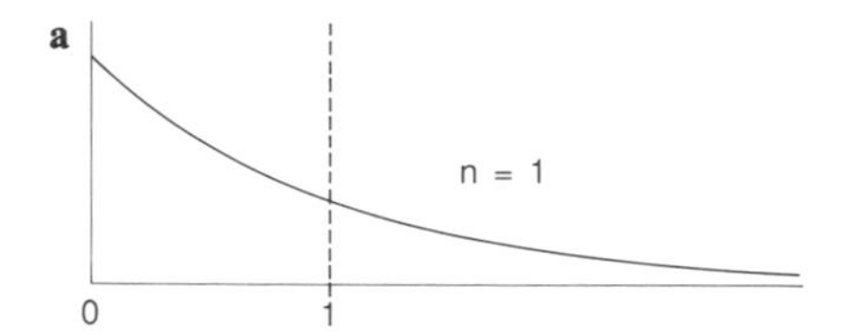

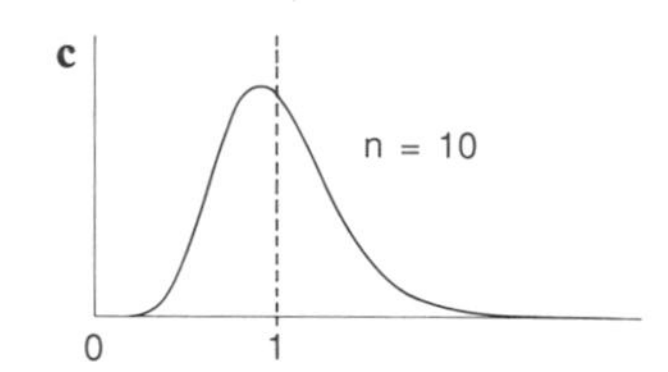

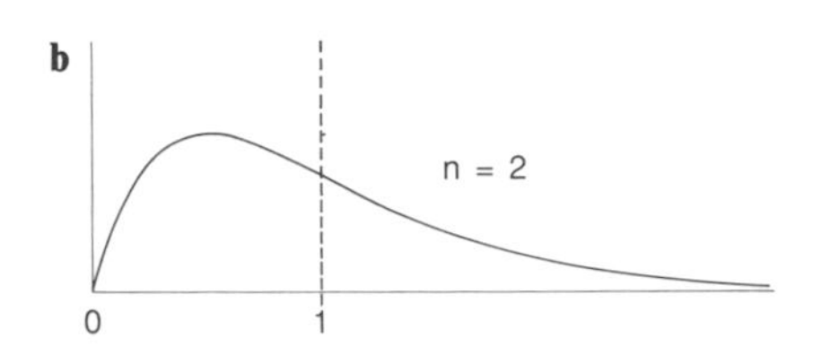

FIGURE 8. The central limit theorem in action: the distribution of means of samples drawn from a skewed distribution (a) displays a progression toward the normal distribution as the sample size increases from 2 (b) to 10 (c).

This is a substantial body of material that is quite forbidding if formally presented. Traditional college instruction in statistics insists that a substantial dose of probability—at least topics on independence and on random variables—precede the study of inference. Some understanding of independence and of distributions with their means and standard deviations is certainly needed. But the degree of mathematical formalism with which these topics are traditionally taught is generally unnecessary at the college level and out of the question in secondary school. Both the length and the difficulty of the path to statistics via formal probability argue against this traditional approach. As Garfield and Ahlgren conclude,[5]

> ... Teaching a conceptual grasp of probability still appears to be a very difficult task, fraught with ambiguity and illusion. Accordingly, we make the pragmatic recommendation for two research efforts that would proceed in parallel: one that continues to explore means to induce valid conceptions of probability, and one that explores how useful ideas of statistical inference can be taught independently of technically correct probability.

Fortunately, the empirical emphasis of data analysis, developed gradually beginning in the early grades, offers a setting for teaching both basic probability and elementary inference. Simulation, first physical and then using software, can demonstrate the essential concepts of probability and is particularly suited to displaying sampling distributions. Only quite informal probability is needed to think about sampling distributions. As the earlier discussion of normal distributions indicated, data description provides an adequate context for distributions as idealized mathematical models for variation. The core mathematics curriculum taught to all students should include data analysis and an empirical introduction to only basic probability concepts and laws at about the level of the Quantitative Literacy material.[15]

INFERENCE

Statistics is concerned with the gathering, organization, and analysis of data and with inferences from data to the underlying reality. "Inference from data to reality" is a knotty topic indeed, with much room for disagreements of a philosophical nature. It is not surprising that statisticians disagree on the most fruitful approach to inference. Barnett[2] gives a comparative overview of the competing positions.

Bayesian or Classical?

The most important philosophical divide separates Bayesian inference from classical inference. Some understanding of the distinction is essential to wise curriculum decisions. The question of inference in simplest form is how to draw conclusions about a population *parameter* on the basis of *statistics* calculated from a sample. A parameter is a number that describes the population, such as the mean height μ of all American women age 18 to 22. A relevant statistic in this case is the sample mean height $\overline{x}$ of a random sample of young women. For purposes of inference we imagine how $\overline{x}$ would vary in repeated samples from the same population. The sampling distribution of the statistic describes this variation. The sampling distribution reflects the underlying parameter—in this case μ is the mean of the distribution of $\overline{x}$. It is because the sampling distribution depends on the unknown parameter that the statistic carries information about the parameter.

Classical inference is rooted in the concept of probability as long-term regularity and in the corresponding idea that the conclusions of inference are expressed in terms of what would happen in repeated data production. To say that we are "95% confident that μ lies between 64.5 and 64.7 inches" is shorthand for "We got this interval by a method that is correct in 95% of all possible samples." Probability statements in classical inference apply to the method rather than to the specific conclusion at hand—indeed, probability statements about a specific conclusion make no sense because the population parameter is fixed, though unknown.

The Bayesian approach wishes to bring prior information about the value of the parameter into play. This is done by regarding the parameter μ as a random quantity with a known probability distribution that expresses our imprecise information about its value. The mean height μ of all American young women is not random in the traditional sense. But it is uncertain. I am quite sure that μ lies between 54 inches and 72 inches, and I think it more likely that μ lies near the center of this range. My subjective assessment of uncertainty can be expressed in a probability distribution for μ.

In the Bayesian view the concept of probability is expanded to include such personal or *subjective probabilities.* What is new here is not the mathematics, which remains the same, but the interpretation of probability as representing a subjective assessment of uncertainty rather than a long-run relative frequency. The sampling distribution of the statistic $\overline{x}$ is now understood to state conditional probabilities of the values of $\overline{x}$ given a value for μ. A calculation then combines the prior information with the observed data to obtain the conditional distribution of μ given the data. (The discrete form of this calculation uses a simple result about conditional probabilities known as Bayes' theorem, from which the Bayesian school takes its name.) The conclusions of inference are expressed in terms of probability statements about the unknown parameter itself: the probability is 95% that the true mean lies between 64.5 and 64.7 inches.

The Bayesian conclusion is certainly easier to grasp than the classical statement. Moreover, prior information is important in many problems. Statisticians generally agree that Bayesian methods should be used when the prior probability distribution of the parameter is known. What is disputed is whether usable prior distributions are always available, as Bayesians contend. Non-Bayesian statisticians do not think that my subjective assessment is always useful information and so are not willing to make general use of subjective prior distributions. The apparently clear conclusion of a Bayesian analysis can depend strongly on assumptions about the prior distribution that cannot be checked from the data.

For introductory instruction about inference, Bayesian methods have several disadvantages. They require a firm grasp of conditional probability. Indeed, students must understand the distinction between the conditional distribution of the statistic given the parameter and the conditional distribution of the parameter given the actually observed value of the statistic. This is fatally subtle. The subjective interpretation of probability is quite natural, but it diverts attention from randomness and chance as observed phenomena in the world whose patterns can be described mathematically. An understanding of the behavior of random phenomena is an important goal of teaching about data and chance; probability understood as personal assessment of uncertainty is at best irrelevant to achieving this goal. The line from data analysis through randomized designs for data production and probability to inference is clearer when classical inference is the goal.

Two types of inference, confidence intervals and significance tests, figure in introductory instruction in classical statistical inference. The reasoning behind both types of inference can be introduced informally in discussions about data. Formal treatment and specific methods should be reserved for upper-grade secondary courses in probability and statistics, and no attempt should be made to present more than a few specific procedures. Particularly in the case of significance tests, a formal approach obscures the reasoning to such an extent that it may be better to avoid hypotheses and test statistics altogether.

Confidence Intervals

The reasoning behind confidence statements is relatively straightforward. What is more, news reports of opinion polls and their margins of error provide a steady supply of examples for discussion. How is it that a sample of only 1500 people can accurately represent the opinion of 185 million American adults? Random sampling provides a part of the explanation; sampling distributions provide the rest, and confidence intervals explain what the margin of error means.

Confidence statements can be introduced after students have some experience with simulation of sampling distributions. The distinction between population and sample, the idea of random sampling, and the notion of a sampling distribution are fundamental to inference. Simulation allows the gradual introduction of confidence intervals during the exploration of sampling and sampling distributions. The ideas of confidence intervals can be taught via graphical display of simulated samples.[10] A more formal approach requires familiarity with normal distributions.

Suppose that in a large county 30% of high school students drive cars to school. Asking a simple random sample of 250 students whether they drove to school today produces 250 independent observations, each with probability 0.3 of being "yes." The proportion $\hat{p}$ of "yes" responses in the sample varies from sample to sample. Simulate (say) 1000 samples and display the sampling distribution of $\hat{p}$. It is approximately normal, with mean 0.3 and standard deviation 0.029. Repeated simulations of samples of various sizes from this population demonstrate that the center of the sampling distribution remains at 0.3 and that the spread is controlled by the size of the sample. In large samples (about 1000 or so) the values of the sample statistic $\hat{p}$ are tightly concentrated around the population parameter $p = 0.3$. Students can see empirically that samples of this size allow good guesses about the entire population.

But just *how* good are guesses based on a sample? We can quantify the answer by describing how the statistic $\hat{p}$ varies in repeated sampling. It is a basic fact of normal distributions that about 95% of all observations lie within two standard deviations on either side of the mean. So in repeated sampling, 95% of all samples of 250 students give a sample proportion $\hat{p}$ within about 0.06 of the true proportion 0.3 who drive to school. The simulation shows that this is so.

Now suppose that a sample of 250 students in another large county finds 105 who drive to school. We guess that the true proportion p of all students in this county who drive to school is close to $\hat{p} = 105/250 = 0.42$. If (as is true) the variability is about the same as in the county we simulated, $\hat{p}$ lies within ± 0.06 of p in 95% of all samples. We say we are *95% confident* that the unknown population proportion p lies in the interval 0.42 ± 0.06. More generally, the interval $\hat{p} \pm 0.06$ is a 95% *confidence interval* for the unknown p.

Figure 9 illustrates the behavior of a confidence interval in repeated samples. As repeated samples of size 250 are drawn, some of the intervals $\hat{p} \pm 0.06$ cover the true proportion of p, while others do not. But in the long run, 95% of all samples produce an interval covering the true p. That is, the probability that the random interval $\hat{p} \pm 0.06$ contains p is 0.95. As is generally the case in classical inference, this probability refers to the performance of the method in an indefinitely large number of repeated samples.

The first portion of the argument above belongs to the study of sampling and simulation and is essentially an empirical demonstration of the surprising trustworthiness of samples that seem small relative to the size of the population. The facts that emerge from such sampling demonstrations are much more important than the formal dress we give them in the second stage of the argument. The second stage belongs to a more advanced study of inference. The qualitative conclusion that most sample results lie close to the truth is made quantitative by giving an interval and a level of confidence. The nature of this conclusion and its limitations both need emphasis.

What are the grounds of our confidence statement? There are only two possibilities.

1. The interval 0.42 ± 0.06 contains the true population proportion p.
2. Our simple random sample was one of the few samples for which $\hat{p}$ is not within 0.06 points of the true p. Only 5% of all samples give such inaccurate results.

We cannot know whether our sample is one of the 95% for which the interval catches p or one of the unlucky 5%. The statement that we are 95% confident that the unknown p lies in 0.42 ± 0.06 is shorthand for "We got these numbers by a method that gives correct results 95% of the time."

As for the limitations on this reasoning, remember that the margin of error in a confidence interval includes only random sampling error.

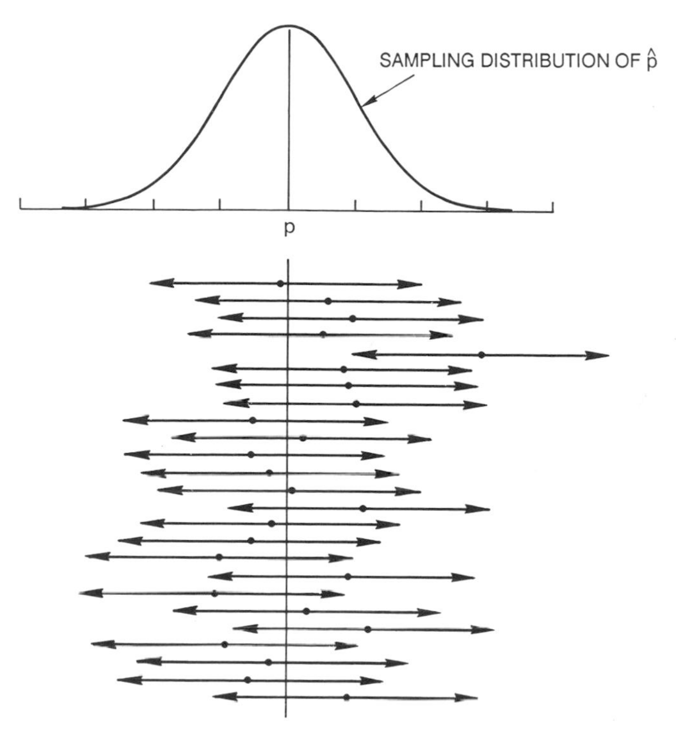

FIGURE 9. The behavior of a confidence interval in repeated samples from the same population. The normal curve is the sampling distribution of the sample proportion $\hat{p}$ centered at the population proportion p. The dots are the values of $\hat{p}$ from 25 samples, with the confidence interval indicated by arrows on either side. In the long run 95% of these intervals will contain p.

In practice there are other sources of error that are not accounted for. For example, national opinion polls are usually conducted by telephone using equipment that dials residential telephone numbers at random. Telephone surveys omit households without phones. Moreover, pollsters often find that as many as 70% of the persons who answer the phone are women. Men will be underrepresented in the sample unless steps are taken to contact males. These facts of real statistical life introduce some bias into opinion polls and other sample surveys.

Significance Tests

The purpose of a confidence interval is to estimate a population parameter and to accompany the estimate with an indication of the uncertainty due to chance variation in the data. Significance tests do not provide an estimate of an unknown parameter, but only an assessment of whether an effect or difference is present in the population. The mere recognition that such an assessment is needed, that not all observed outcomes signify a real underlying cause, already shows statistical sophistication. Judges of science fair displays who talk to the able students who have prepared them find that any effect in the desired direction, however small, is regarded as convincing. The role of chance variation is not recognized.

Statistical significance is a way of answering the question "Is the observed effect larger than can reasonably be attributed to chance alone?" Here is the reasoning of significance tests presented informally in the setting of an important example:

> During the Vietnam era, Congress decided that young men should be chosen at random for service in the army. The first draft lottery was held in 1970. Birth dates were drawn in random order and men were drafted in the order in which their birth dates were selected. After the drawing, news organizations claimed that men born late in the year were more likely to get low draft numbers and so to be inducted. Data analysis (Figure 10) does suggest an association between birth date and draft number. A statistic that measures the strength of the association between draft number (1 to 366) and birth date (1 to 366 beginning with January 1) is the correlation coefficient. In fact, $r = -0.226$ for the 1970 lottery. Is this good evidence that the lottery was not truly random?

A significance test approaches the issue by asking a *probability question:* Suppose for the sake of argument that the lottery were truly random; what is the probability that a random lottery would produce an r at least as far from 0 as the observed $r = -0.226$? *Answer:* The probability that a random lottery will produce an r this far from 0 is less than 0.001. *Conclusion:* Since an r as far from 0 as that observed in 1970 would almost never occur in a random lottery, we have strong evidence that the 1970 lottery was not random.

Figure 10 displays the scatterplot of draft numbers assigned to each birth date by the 1970 draft lottery. It is difficult to see any systematic association between birth date and lottery number in the scatterplot. Clever graphics can emphasize the association, as in the figure. But a probability calculation is needed to learn whether the observed association is larger than can reasonably be attributed to chance alone.

In a random assignment of draft numbers to birth dates, we would expect the correlation to be close to 0. The observed correlation for the

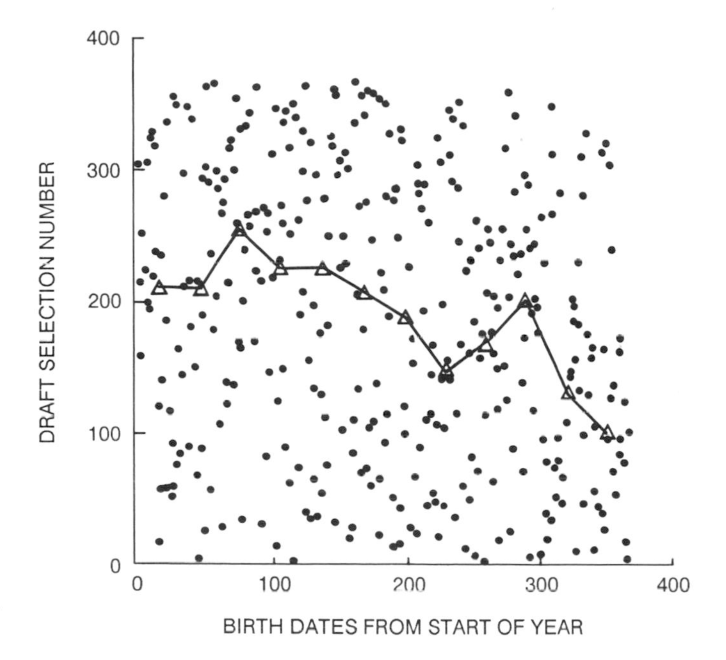

FIGURE 10. Data from the 1970 draft lottery reveal a slight negative correlation, with birth dates near the end of the year most likely to have low draft numbers. The trend can be seen more readily by plotting the median draft numbers for each month. The plot of monthly medians connected by line segments to display the trend, called a *median trace*, is a common tool used to highlight patterns in scatterplots of a response variable against an explanatory variable.

1970 lottery was $r = -0.226$, showing that men born later in the year tended to get lower draft numbers. Common sense alone cannot decide if $r = -0.226$ means that the lottery was not random. After all, the correlation in a random lottery will almost never be exactly 0. Perhaps that $r = -0.226$ is within the range of values that could plausibly occur due to chance variation alone.

To resolve this uncertainty we compare the observed -0.226 to a reference distribution, the sampling distribution of r in a truly random lottery. We find that a truly random lottery would almost never produce an r as far from zero as the r observed in 1970. The probability calculation tells us what common sense could not—that $r = -0.226$ is a large effect, a surprising effect in a random lottery. This convinces us that the 1970 lottery was biased. Investigation disclosed that the capsules containing the birth dates had been filled a month at a time and not adequately mixed. Later dates remained near the top and tended to be drawn earlier. (Fienberg[4] gives more detail about the 1970 draft lottery, including extensive statistical analysis of the outcome.)

Questions like "Is this a large outcome?" or "Is this a surprising result?" come up often in analyzing data. It is quite natural to give an answer by comparing the individual outcome to a reference distribution, as we informally compare the birth weight of a child to the distribution of birth weights of all children. Students should certainly be encouraged to recognize the role of chance variation and to assess "significance" informally by comparing an individual outcome to a suitable reference distribution. If probability and computer simulation are being developed, the comparison can be put in the language of probability and sampling distributions. But formal "tests of hypotheses" need not appear in the school curriculum.

There are several reasons for this. The mechanics of stating hypotheses, calculating a test statistic, and comparing with tabled values effectively conceal the reasoning of significance tests. The reasoning itself is somewhat difficult and full of subtleties. Effective examples of the use of significance tests are more removed from everyday experience than opinion polls and similar examples of confidence statements. An understanding of data and chance, and the development of quantitative reasoning in general, is better served by concluding the study of statistics in the schools with probability, sampling distributions, confidence intervals, and a continuing emphasis on using these tools in reasoning about uncertain data.

STATISTICAL THINKING

Statistics and probability are the sciences that deal with uncertainty, with variation in natural and man-made processes of every kind. As such they are more than simply a part of the mathematics curriculum, although they fit well in that setting. Probability is a field within mathematics. Statistics, like physics or economics, is an independent discipline that makes heavy and essential use of mathematics.

Statistics has some claim to being a fundamental method of inquiry, a general way of thinking that is more important than any of the specific facts or techniques that make up the discipline. If the purpose of education is to develop broad intellectual skills, statistics merits an essential place in teaching and learning. Education should introduce students to literary and historical methods; to the political and social analysis of human societies; to the probing of nature by experimental science; and to the power of abstraction and deduction in mathematics. Reasoning from uncertain empirical data is a similarly powerful and pervasive intellectual method.

This is not to say that detailed instruction in specific statistical methods for their own sake should be prominent in the school curriculum. Indeed, they should not. But statistical thinking, broadly understood, should be part of the mental equipment of every educated person. We can summarize the core elements of statistical thinking as follows:

1. The omnipresence of *variation* in processes. Individuals are variable; repeated measurements on the same individual are variable. The domain of a strict determinism in nature and in human affairs is quite circumscribed.
2. The need for *data* about processes. Statistics is steadfastly empirical rather than speculative. Looking at the data has first priority.
3. The design of *data production* with variation in mind. Aware of sources of uncontrolled variation, we avoid self-selected samples and insist on comparison in experimental studies. And we introduce planned variation into data production by use of randomization.
4. The *quantification* of variation. Random variation is described mathematically by *probability.*
5. The *explanation* of variation. Statistical analysis seeks the systematic effects behind the random variability of individuals and measurements.

Statistical thinking is not recondite or removed from everyday experience. But it will not be developed in children if it is not present in the curriculum. Students who begin their education with spelling and multiplication expect the world to be deterministic; they learn quickly to expect one answer to be right and others wrong, at least when the answers take numerical form. Variation is unexpected and uncomfortable. Listen to Arthur Nielsen[16] describing the experience of his market research firm with sophisticated marketing managers:

> ...Too many business people assign equal validity to all numbers printed on paper. They accept numbers as representing Truth and find it difficult to work with the concept of probability. They do not see a number as a kind of shorthand for a range that describes our actual knowledge of the underlying condition. For example, the Nielsen Company supplies to manufacturers estimates of sales through retail stores. ... I once decided that we would draw all charts to show a probable range around the number reported; for example, sales are either up 3 percent or down 3 percent or somewhere in between. This turned out to be one of my dumber ideas. Our clients just couldn't work with this type of uncertainty. They act as if the number reported is gospel.

The ability to deal intelligently with variation and uncertainty is the goal of instruction about data and chance. There is some evidence that instruction actually improves this ability. Nisbett et al.[17] describe research on teaching various kinds of reasoning. They note that instruction in probability and statistics increases the willingness to consider chance variation even when the instruction is of a traditional kind that makes no attempt to apply probabilistic reasoning in unstructured settings. Here is a typical example:

> [Subjects were asked] to explain why a traveling saleswoman is typically disappointed on repeat visits to a restaurant where she experienced a truly outstanding meal on her first visit. Subjects who had no background in statistics almost always answered this problem with exclusively nonstatistical, causal answers such as "maybe the chefs change a lot" or "her expectations were so high that the food couldn't live up to them." Subjects who had taken one statistics course gave answers that included statistical considerations, such as "very few restaurants have only excellent meals, odds are she was just lucky the first time," about 20 percent of the time.

Nisbett and his colleagues find it striking that instruction of a quite abstract kind does have an effect on thinking about everyday occurrences. The effect is stronger when instruction points out the applicability of statistical ideas in everyday life, as school instruction should certainly do. This is evidence that we are in fact dealing with a *fundamental* and generally applicable intellectual skill. Nisbett also reports research showing that training in deterministic disciplines, even at the graduate level, does not similarly improve everyday statistical reasoning. This is evidence that we are dealing with an *independent* intellectual method.

Why teach about data and chance? Statistics and probability are useful in practice. Data analysis in particular helps the learning of basic mathematics. But, most important, it is because statistical thinking is an independent and fundamental intellectual method that it deserves attention in the school curriculum.

REFERENCES AND RECOMMENDED READING

1. BBN Laboratories. *ELASTIC and Reasoning Under Uncertainty.* Final Report, 1989, p. 30.
2. Barnett, Vic. *Comparative Statistical Inference, Second Ed.* New York, NY: John Wiley & Sons, 1982.
3. Efron, Bradley. "Computers and the theory of statistics: Thinking the unthinkable." *SIAM Review,* 21 (1979), 419–437.
4. Fienberg, Stephen E. "Randomization and social affairs: The 1970 draft lottery." *Science,* 171 (1971), 255–261.
5. Garfield, Joan and Ahlgren, Andrew. "Difficulties in learning basic concepts in probability and statistics: Implications for research." *Journal for Research in Mathematics Education,* 19 (1988), 44–63.
6. Gastwirth, Joseph. "The statistical precision of medical screening procedures: Application to polygraph and AIDS antibodies test data." *Statistical Science,* 2 (1987), 213–238.

7. Gnanadesikan, Mrudulla; Schaeffer, Richard; Swift, James. *The Art and Techniques of Simulation.* Palo Alto, CA: Dale Seymour Publishers, 1986.
8. Jones, L.V. (Ed.). *The Collected Works of John W. Tukey. Vol. 3: Philosophy and Principles of Data Analysis, 1949–1964; Vol. 4: Philosophy and Principles of Data Analysis, 1965–1986.* Monterey, CA: Wadsworth & Brooks/Cole, 1986.
9. Kruskal, William. "Miracles and statistics: The casual assumption of independence." *Journal of the American Statistical Association,* 83 (1988), 929–940.
10. Landwehr, James and Watkins, Ann. *Exploring Data.* Palo Alto, CA: Dale Seymour Publishers, 1986.
11. Landwehr, James; Watkins, Ann; Swift, James. *Exploring Surveys and Information from Samples.* Palo Alto, CA: Dale Seymour Publishers, 1987.
12. Mathematical Sciences Education Board. *Reshaping School Mathematics: A Philosophy and Framework for Curriculum.* National Research Council. Washington, DC: National Academy Press, 1990.
13. Moore, David and McCabe, G. *Introduction to the Practice of Statistics.* New York, NY: W.H. Freeman, 1989.
14. National Council of Teachers of Mathematics. *Curriculum and Evaluation Standards for School Mathematics.* Reston, VA: National Council of Teachers of Mathematics, 1989.
15. Newman, Claire; Obremski, Thomas; Schaeffer, Richard. *Exploring Probability.* Palo Alto, CA: Dale Seymour Publishers, 1986.
16. Nielsen, Arthur C., Jr. "Statistics in marketing." In Easton, G.; Roberts, Harry V.; Tiao, George C. (Eds.): *Making Statistics More Effective in Schools of Business.* Chicago, IL: University of Chicago Graduate School of Business, 1986.
17. Nisbett, Richard E.; Fong, Geoffrey T.; Lehman, Darrin R.; Cheng, Patricia W. "Teaching reasoning." *Science,* 238 (1987), 625–631.
18. Rubin, Andee; Bruce, Bertram; Rosebery, Ann; DuMouchel, William. "Getting an early start: Using interactive graphics to teach statistical concepts in high school." *Proceedings of the Statistical Education Section.* American Statistical Association, 1988, 6–15.
19. Rubin, Andee and Rosebery, Ann. "Teachers' misunderstandings in statistical reasoning: Evidence from a field test of innovative materials." In Hawkins, Ann (Ed.): *Training Teachers to Teach Statistics.* Proceedings of an International Statistics Institute Roundtable, July 1988.
20. Tversky, Amos and Kahneman, Daniel. "Belief in the law of small numbers." *Psychological Bulletin,* 76 (1971), 105–110.
21. Tversky, Amos and Kahneman, Daniel. "Extensional versus intuitive reasoning: The conjunction fallacy in probability judgment." *Psychological Review,* 90 (1983), 293–315.
22. Vallone, Richard and Tversky, Amos. "The hot hand in basketball: On the misperception of random sequences." *Cognitive Psychology,* 17 (1985), 295–314.

Shape

~ ~

MARJORIE SENECHAL

INTRODUCTION

We encounter patterns all the time, every day: in the spoken and written word, in musical forms and video images, in ornamental design and natural geometry, in traffic patterns, and in objects we build. Our ability to recognize, interpret, and create patterns is the key to dealing with the world around us.

Shapes are patterns. Some shapes are visual, evident to everyone: houses, snowflakes, cloverleafs, knots, crystals, shadows, plants. Others, like eight-dimensional kaleidoscopes or four-dimensional manifolds, are highly abstract and accessible to very few.

"The increasing popularity of puzzles and games based on the interplay of shapes and positions illustrates the attraction that geometric forms and their relations hold for many people," observed geometer Branko Grünbaum. "Patterns are evident in the simple repetition of a sound, a motion, or a geometric figure, as in the intricate assemblies of molecules into crystals, of cells into higher forms of life, or in countless other examples of organizational hierarchies. Geometric patterns can serve as relatively simple models of many kinds of phenomena, and their study is possible and desirable at all levels."

But despite their fundamental importance, students learn very little about shapes in school. The study of shape has historically been subsumed under geometry (literally "earth measurement"), which for a long time has been dominated by postulates, axioms, and theorems of Euclid.

Just as Shakespeare is not sufficient for literature and Copernicus is not sufficient for astronomy, so Euclid is not sufficient for geometry. Like scholars in all times and places, Euclid wrote about the concepts of geometry that he knew and that he could treat with the methods available to him. Thus he did *not* write about the geometry of maps, networks, or flexible forms, all of which are of central importance today.

Shape is a vital, growing, and fascinating theme in mathematics with deep ties to classical geometry but goes far beyond it in content, meaning, and method. Properly developed, the study of shape can form a central component of mathematics education, a component that draws on and contributes to not only mathematics but also the sciences and the arts.

Like many other important concepts, "shape" is an undefinable term. We cannot say precisely what "shape" means, partly because new kinds of shapes are always being discovered. We assume we know what shapes are, more or less: we know one when we see one, whether we see it with our eyes or in our imaginations.

But we know much more than this. We know that shapes may be alike in some ways and different in others. A football is not a basketball, but both are smooth closed surfaces; a triangle is not a square, but both are polygons. We know that shapes may have different properties: a triangle made of straws is rigid, but a square made of straws is not. We know that shapes can change and yet be in some way the same: our shadows are always *our* shadows, even though they change in size and contour throughout the day.

In the study of shape, our goals are not so very different from those of the ancient Greek philosophers: to discover similarities and differences among objects, to analyze the components of form, and to recognize shapes in different representations. Classification, analysis, and representation are our three principal tools. Of course, these tools are closely interrelated, so distinctions among them are to some extent artificial. Is symmetry a tool for classifying patterns or a tool for analyzing them? In fact, it is both. Nevertheless, it is helpful to discuss each of these tools separately.

CLASSIFICATION

One of the great achievements of ancient mathematics was the discovery that there are exactly five convex, three-dimensional shapes whose surfaces are composed of regular polygons, with the same number of polygons meeting at each corner. These shapes, known as the *regular polyhedra,* are shown in Figure 1. This discovery so excited the imagination of the ancients that Plato made these shapes the cornerstone of his theory of matter (see his dialogue *Timaeus*), and Euclid devoted

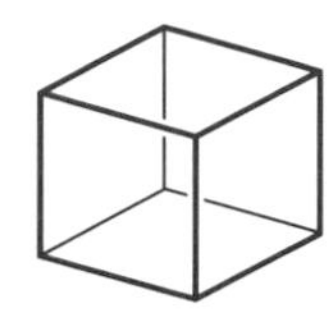

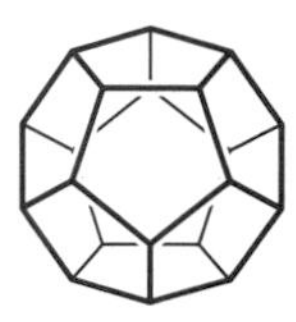

FIGURE 1. The five regular polyhedra. Each is composed of a single type of regular polygon, with the same number of polygons meeting at each corner. The tetrahedron, octahedron, and icosahedron are made of triangles, the cube is made of squares, and the dodecahedron is made of pentagons.

much of Book XIII of his *Elements* to their construction. They have lost none of their fascination today.

It is easy today to underestimate the significance of the discovery of the regular polyhedra. In its time it was a major feat of mathematical imagination. In the first place, in order to count the number of objects of a certain kind you have to be aware that they are "of a certain kind." That is, you must recognize that these objects have properties that distinguish them from other objects and be able to characterize their distinguishing features in an unambiguous way. Second, you must be able to use these criteria to find out precisely which objects satisfy them. No one knows just how the ancients made their discovery, but it is easy for young children today, especially if they have regular polygons to play with, to convince themselves that the list of regular polyhedra is complete (Figure 2).

The key ingredients of mathematical classification were already in use thousands of years ago: characterizing a class of objects and enumerating the objects in that class. What has changed throughout the centuries, and will continue to change, are the kinds of characterizations that seem important to us and the methods that we use for enumeration. Figure 3 shows several classes of objects that can be grouped together from a mathematical point of view. Examples such as these can stimulate student discussion: What properties characterize each class? Are there different ways to classify these objects? What other objects belong to these classes? We mention here a few of the classification schemes that have proved effective in many applications.

Congruence and similarity. Two objects are congruent if they are exactly alike down to the last detail, except for their position in space. Cans of tomato soup (of the same brand) in a grocery store, square tiles on a floor, and hexagons in a quilt pattern are all familiar examples of congruent figures. Two objects are similar if they differ only in position and scale. Similarity seems to be a very fundamental concept. Preschoolers understand that miniature animals, doll clothes, and play houses are all small versions of familiar things. The fact that even such young children know what these tiny objects are supposed to represent

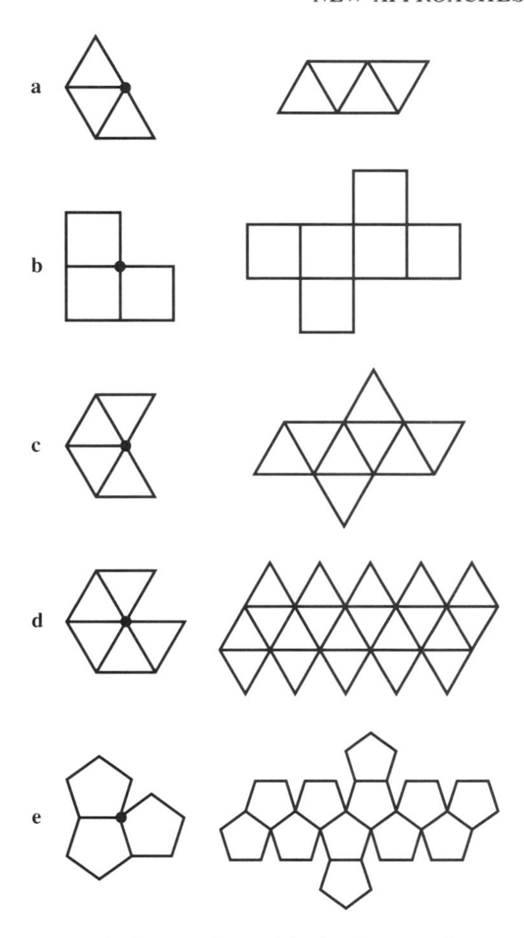

FIGURE 2. There are only five regular polyhedra because there are only five arrangements of congruent, regular polygons about a point that can be folded up to make a convex polygonal vertex. Here we see the five arrangements, together with their completion as patterns that can be folded up to make the entire polyhedron.

shows that they intuitively understand change of scale. Building and taking apart scale models of towers, bridges, houses, shapes of any kind give the child—of any age—a firm grasp of this idea.

Symmetry and self-similarity. A square is symmetrical: if you rotate it 90°, 180°, 270°, or 360° about its center, it appears unchanged. Also, it has four lines of mirror symmetry across which you can reflect it onto itself (Figure 4). It is easy to think of other objects that have the same symmetries, or self-congruences, as the square: the Red Cross symbol, a bracelet with four equally spaced beads, a circle of four dancers, and a four-leaf clover (without its stem) are a few examples. Symmetry classifies objects according to the arrangement of their constituent parts.

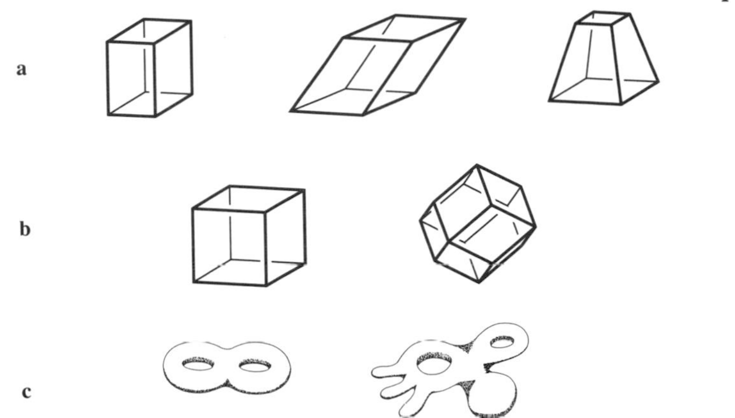

FIGURE 3. Examples of solid objects grouped into useful classes. What do the shapes in each class have in common?

This can be rather subtle; for example, the two polyhedra in Figure 3b have the same symmetries.

Just as congruence leads to symmetry (which is just another name for self-congruence), so similarity extends naturally to self-similarity. "Thc basic fact of aesthetic experience," according to art historian E.H. Gombrich,[9] "is that delight lies somewhere between boredom and confusion." Perhaps this is one of the reasons why fractals and other self-similar figures are generating so much excitement.

"Beauty is truth, truth beauty," wrote the poet John Keats. Self-similarity has recently been recognized as a profound concept in nature. The awarding of a Nobel prize for the formulation of "renormalization groups" and the current worldwide cross-disciplinary interest in chaos theory indicate the profound implications of similarity and scale for science and mathematics. The study of scaling has stimulated (and been stimulated by) the study of fractals and other self-similar geometrical forms.

Combinatorial properties. Congruence and similarity are metric concepts: they can be altered by changing lengths or angles. But some other

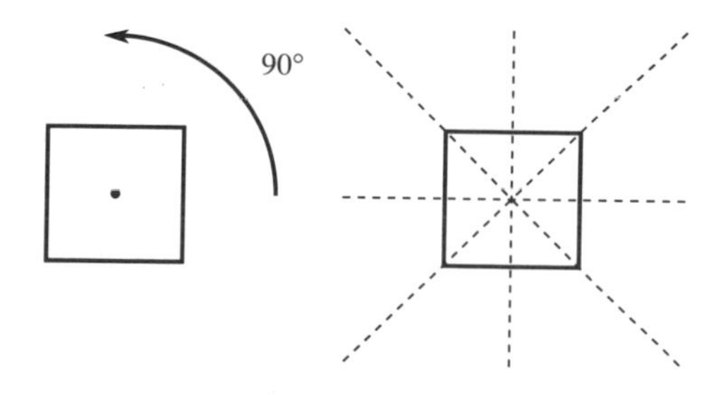

FIGURE 4. If a square is rotated 90°, 180°, 270°, or 360° about its center, it appears unchanged. Also, it has four lines of mirror symmetry across which you can reflect it onto itself.

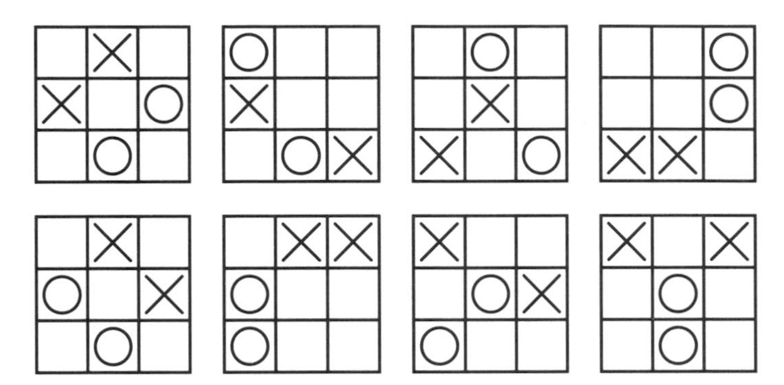

FIGURE 5. In torus tic-tac-toe the opposite sides of the board are identified—that is, considered to be the same. It is as if the board were rolled into a cylinder, which was then bent around to form an inner-tube shape that mathematicians call a torus. Can you tell which of these positions are equivalent in the torus-shaped game?

properties of shapes remain intact under such transformations. For example, the numbers of edges and vertices of a polygon are not altered if we stretch or bend the polygon. Thus the three hexagons of Figure 7 are all hexagons, even though they are neither congruent nor similar: a hexagon is any closed loop made of six line segments. Being a hexagon is a combinatorial property of a polygon.

Roughly speaking, the combinatorial properties of a shape are the things we can count and the way they are fitted together. Thus from the combinatorial point of view, the shapes in Figure 3a are equivalent, since each has 6 faces, 8 vertices, and 12 edges connected to each other in the same way. Network problems often involve combinatorial problems. For example, if we want to design a linking system for the computers in a building, we are concerned first with finding the possible arrangements of links and nodes that can provide the connections we want, and only then need we consider how long the cables will have to be.

Topology. Topological equivalence is even more general than combinatorial equivalence. From the standpoint of topology, all polygons are loops and all convex polyhedra are alike. Piaget argued that topological concepts occur prior to metric ones in child development; a child may recognize a loop before distinguishing among kinds of loops, such as circles and triangles. Being a loop, as opposed to a knot, is a topological property of shape.

Topology in school is often described as "rubber sheet geometry." It yields many excellent examples that can enlarge a child's concept of the flexibility of shape. In rubber sheet geometry, the shapes of Figure 3c are indistinguishable because each can be deformed into the other. Knots, of the boy and girl scout variety, are an excellent subject for hands-on study.[12] Children can learn to play tic-tac-toe on a torus and other delightful games that require geometrical mental gymnastics (Figure 5).

With complexity of structure, topological classification necessarily becomes more sophisticated. Here computer visualization can be a useful tool. Older students can appreciate the concept of orientation, which characterizes the difference between a cylinder and a Mobius band (orientable and non-orientable), and the concept of genus, which characterizes the topological difference between a sphere and a torus (genus zero and genus one). Understanding such concepts enriches greatly the study of science and design as well as mathematics.

Naming

Shapes need names. One of the most fundamental uses of language is to assign names to things. Naming is a primitive concept that is echoed in our myths as well as in many contemporary religious practices. Naming is the first step toward knowing, whether it is the name of a person or the name of a shape. We cannot think about shapes (or anything else for that matter) or explain our ideas to others if we do not use names. Learning technical names is sometimes disparaged as a rote activity, but such objections miss the point. Technical names are usually not arbitrary; they encode the conceptual framework in which we organize the things we are naming.

For example, in English-speaking countries, last names indicate the family and first names designate an individual in a family. Thus Mary Jones is a person named Mary who is a member of the Jones family. The names of shapes serve similar functions: a tetrahedron is a member of the polyhedron family, a representative of the subfamily of those polyhedra that have four faces (see Figure 6). When we use the word

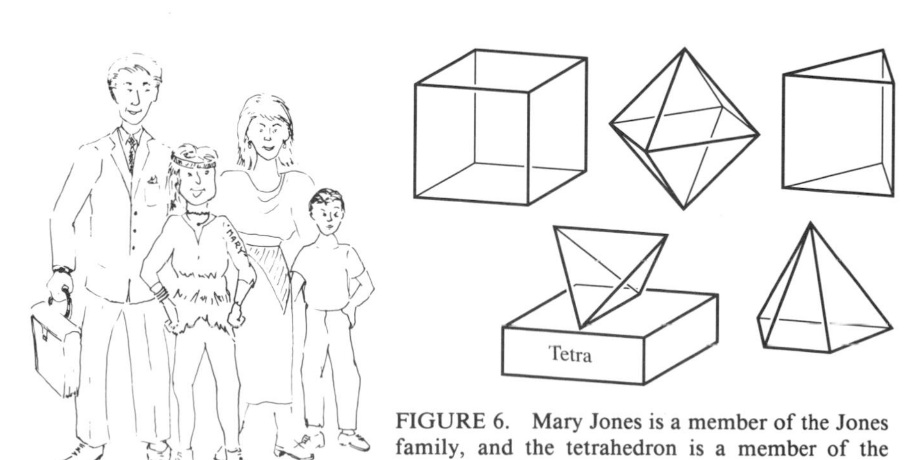

FIGURE 6. Mary Jones is a member of the Jones family, and the tetrahedron is a member of the polyhedron family.

"tetrahedron" to name a shape, we are at the same time locating it in its family tree and describing it in a meaningful way.

Although classification requires precision, there is no single "right" way to classify shapes. Shapes are classified into families and subfamilies in many different ways, depending on the properties that interest us. For example, the discovery that the orbits of the planets around the sun are ellipses, and not circles, revolutionized the study of astronomy; from this standpoint circles and ellipses are completely different. But one of the great achievements of the ancients was the discovery that both circles and ellipses are conic sections and in that sense are the same.

From the point of view of topology, the distinction between shapes that enclose regions, like balls, and shapes that have holes in them, like bagels, is fundamental; within these broad classes, all shapes are alike. But a football player would not be happy with a basketball as a substitute, nor would a basketball player be willing to make do with a baseball, because the individual kinds of balls have crucially different properties. As another example of cross-classification, architects know that it is important to build houses that are sturdy, not houses that might collapse. This concern transcends other ways that houses are commonly classified, such as large and small, single story or multistory, rectangular, or dome-like.

Classification skills develop gradually. Very young children learn to recognize a great many shapes without being formally taught. Their world is literally made of shapes: shapes that hold things, such as bowls and bags and baskets; shapes to play with, such as balls and puzzles and blocks; shapes to use, such as chairs and spoons and beds. Thousands of shapes are part of children's lives. Later, in school, children learn

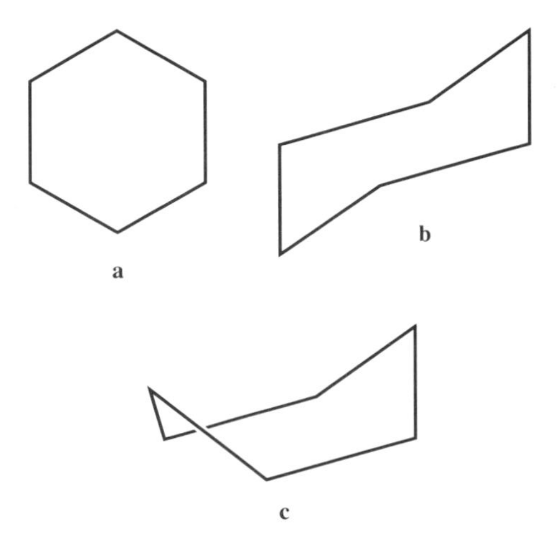

FIGURE 7. Three hexagons that are important in chemistry. The planar hexagon (a) occurs in benzene (see also Figure 30 below). The hexagons in (b) and (c) are intended to be nonplanar; both are conformations of cyclohexane. Hexagons made out of flexible straws can easily assume any of these shapes.

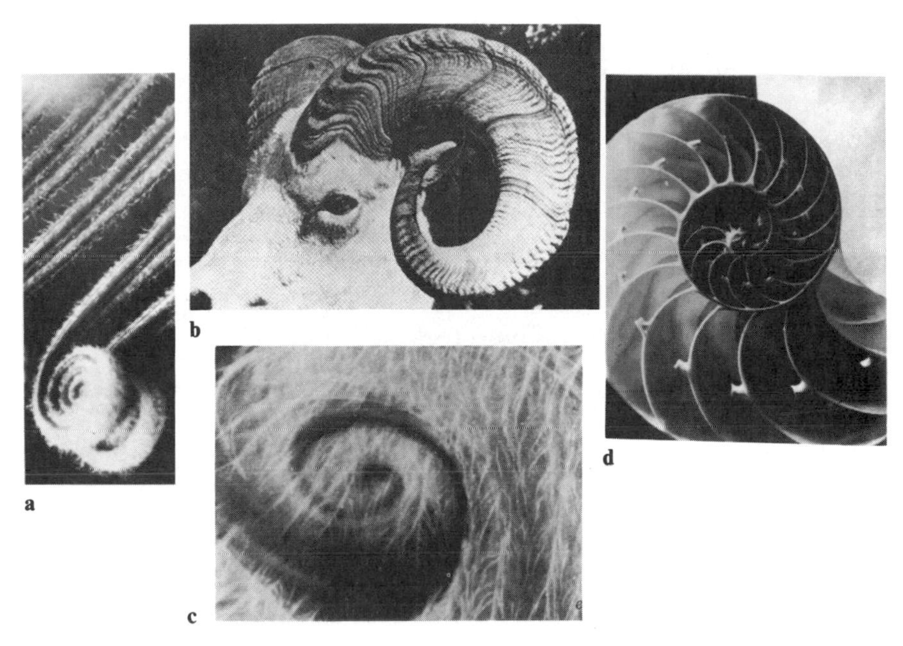

FIGURE 8. Four natural spirals: (a) leaves of the sago palm, (b) horns of a mountain sheep, (c) glycerin mixed with food coloring and ink, (d) the chambered nautilus. The common shape suggests a common creative mechanism, despite the striking differences in material, scale, and natural forces.

names for some of them, such as circles, spheres, polygons, and some simple polyhedra.

Alas, in our schools identification and classification of shapes usually stop just at the point where they can begin to be really interesting—where they begin to explore structures in three-dimensional space. How many people realize that even polygons that are not flat can be interesting and important? Many molecules have polygonal shapes, but often these polygons are crumpled and their conformations are the key to their chemical properties (Figure 7). Besides finite polygons and polygons whose edges don't cross, there are zig-zag, star, and helical polygons. By broadening the definition of polygon to include any closed loop, we may also study knots. In addition to their obvious practical importance for tying things, knots enter into the design of networks such as cloverleafs and are helpful in understanding the structure of some biological molecules. Soap bubbles, soap films, and froths are also endless sources of fascinating geometrical principles.

The study of polyhedra can be extended from simple shapes that are easy to construct to others, such as star polyhedra, that are more complex. Equally important are patterns, such as tilings of the plane, that are beautiful as well as useful. The helix and the spiral are fundamental to biology and astronomy as well as to mathematics. But even today, when "double helix" has become almost a household phrase, few people realize that there is a fundamental difference between a helix, which twists around an axis at a constant distance from it, and a spiral (Figure 8). Most so-called spiral staircases are really helical, for obvious practical reasons. Imagine what we would be like if our DNA wound itself in spirals, or what the universe would be like if galactic spirals were helices!

ANALYSIS

In order to interpret and create patterns in today's image-packed world, it is not enough just to recognize similarities and differences; we also need to analyze them. This leads us to investigate the way that large shapes are built of smaller ones and to recognize patterns and their properties.

When children make shapes out of blocks or Legos, they often imitate the diverse compositions that they see around them (Figure 9). Nature too creates patterns. Like man-made patterns, natural patterns appear at many levels: atoms are organized into molecules, while molecules are organized into crystals and cells, which in turn are often the subunits of still higher-level organization.

When we examine patterns carefully, we find that the same forms and arrangements appear over and over again, even when the objects

FIGURE 9. Many shapes are built from smaller ones. The reinforcing beams in a bridge illustrate how repeated patterns are used in engineering and architecture, as in nature, to form a whole out of parts.

FIGURE 10. Young children can investigate the ways in which polygons can be fitted together to tile a plane surface.

involved are very different.[16] This is not just a coincidence. The geometry of most patterns is governed by a very few basic principles of formation, growth, and development. For example, in his fascinating book *Patterns in Nature*,[20] Peter Stevens discusses several ways in which natural patterns are generated, such as stress, branching, meandering, partitioning, close packing, and cracking. The results of these modes of formation are remarkably similar, despite the variety of materials on which they operate (see Figure 8).

Important aspects of pattern formation can be grasped by exploring the ways in which copies of objects can be packed together. Students quickly discover that there are only a very few ways to do this. This fundamental property of shape can be studied at many levels. For example, it can be studied intuitively and "hands-on" when the objects being packed are circles or easy-to-construct polygons such as triangles, quadrilaterals, and hexagons (Figure 10). Older children can experiment with less regular forms and discover some surprising things, such as the fact that any quadrilateral, even one that is not convex, will tile the plane (Figure 11). (This is a surprising but very simple consequence of the fact that the sum of the measures of the angles of a quadrilateral is 360°.) High school students can study deeper properties of sphere packing and tilings, such as their symmetry and how they can be generated. (Grünbaum and Shephard's *Tilings and Patterns*[10] is the definitive resource for material on tilings.)

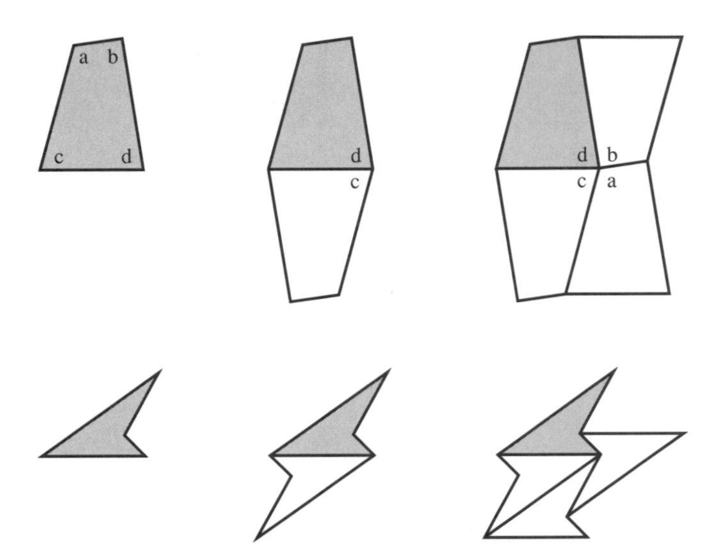

FIGURE 11. Any quadrilateral will tile the plane, because the sum of the measures of its angles is 360°, which is the same as the total number of degrees around each vertex. So four copies of a quadrilateral arranged around a point with each angle used once will fit just perfectly.

Discovering Symmetry

One of the most striking things about patterns of many kinds is their symmetry, and this symmetry is an important tool in their analysis. A pattern is something that repeats in some sense; symmetry is the concept that makes that sense precise.

The study of symmetry begins by decomposing figures into congruent parts. Although some shapes do not at first appear to be made of smaller parts, it is often helpful to think of them as if they were. For example, mirror lines divide a square into eight congruent sectors, which the symmetries of the square permute. This decomposition helps us study the way symmetries work. In particular, it reveals that symmetry is self-congruence. It is this self-congruence that we consider beautiful and that makes symmetry a meaningful organizing principle in the analysis of structure.

Young children learn quite easily to recognize symmetry, not only in squares and butterflies, but also in animals, flowers, household utensils, toys, buildings, and arrays of every kind. Symmetry can be found almost everywhere. Older children can get great pleasure, and gain great insight, by creating symmetrical patterns and discovering the rules that govern them.

One of the most interesting but underappreciated techniques for exploring patterns is paper folding. We are all familiar with the pretty patterns that result when folded paper is cut and then unfolded. The snowflakes, chains of dolls, and other repeating patterns that appear are not created by magic but are simple consequences of the geometry of reflection. Many geometrical constructions, and even aspects of number theory (some of them decidedly nontrivial), can be represented by unfolded designs. Conversely, many interesting three-dimensional shapes can be created by folding paper: the polyhedral nets of Figure 2 are one example; origami puzzles are another. Paper-folding problems stimulate the geometrical imagination in many ways.

Mirror Geometry

Mirrors can be used to study the principles of reflection. In particular, building a kaleidoscope is an excellent way to discover how reflections interact to generate the orderly arrangements that we call kaleidoscopic patterns. The kaleidoscope is much more than a toy: it is a lesson in mirror geometry. Even one mirror has much to teach us: adults as well as children are challenged by the "mirror cards" used in elementary school classes. The kaleidoscope is more complex, but it too is based on the principles of reflection in a mirror.

To explore the operation of a simple kaleidoscope, you just need two rectangular pocket mirrors and some tiny colored objects—bits of plastic or glass will do very well. Tape the two mirrors together along one edge, with their reflecting surfaces facing each other. Place the objects on a table, between the standing mirrors (Figure 12). If you look in the mirrors you will see the objects repeated in a delightful pattern. A little experimentation will show that some angles produce lovelier configurations than others. Only certain angles produce, in the words of the kaleidoscope's inventor, Sir David Brewster, "a perfect whole"—a finite number of identical regions arranged in a circular pattern. By playing with the mirrors, it is not difficult for children to discover which angles produce this perfect kaleidoscopic image. By doing so they will have learned an important lesson in the modern study of shape.

Reflections generate patterns with a finite number of subunits, patterns that have rotational as well as mirror symmetry. The rotations and reflections can be performed one after the other, always leaving the "perfect whole" apparently unchanged. Formally, such a system of motions is known as a *symmetry group.* Many properties of shapes can be analyzed by studying their symmetry groups; indeed, for more than a century this strategy has been a guiding principle in the study of geometry. By using a kaleidoscope, students can understand this fundamental idea by direct experience without making a lengthy detour through the formal and abstract algebraic language in which it is usually expressed.

FIGURE 12. The principle of the kaleidoscope is discovered by playing with two hinged pocket mirrors. The objects appear repeated in infinitely varied patterns, but as the angle between the mirrors is changed, some patterns reveal greater symmetry (and beauty) than others.

FIGURE 13. A pyrite crystal. The lines on the cube's faces indicate that the crystal's internal structure lacks some of the symmetries of the cube.

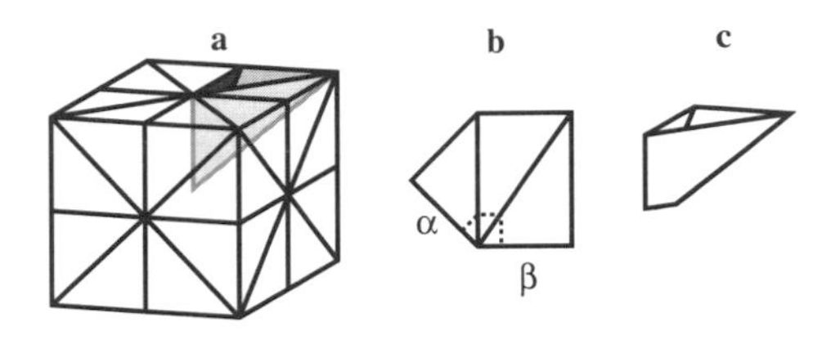

FIGURE 14. A cubic kaleidoscope can be made by placing mirrors or reflecting mylar on the inside of three sides of one of the tetrahedral sectors into which the cube is divided by its mirror planes (a). The net for these three walls is shown in (b); it consists of half a square and a rectangle whose base is the length of the square's edge and whose height is the length of the square's diagonal. Cut along the dotted lines, and then tape the edges α and β together (c). With the cut end down and parallel to a table, look at a piece of newspaper or other decorated material through the tetrahedron. You will see a decorated cube! By moving the tetrahedron along the plane surface, you will see a changing pattern on the cube.

The symmetry of three-dimensional figures appears to be more intricate, but actually the principles are the same as in the two-dimensional case. For example, the symmetry of the cube includes reflections in two kinds of mirror planes and rotations about three kinds of axes. Younger students can learn a great deal about the symmetry of the cube by trying to decorate it in ways consistent with its symmetry. Older students can be challenged by the task of changing this symmetry by decoration.

Such decorations appear in nature, where they provide clues to the structure of hidden patterns. For example, the pyrite crystal in Figure 13 appears at first glance to be an ordinary cube, but closer inspection reveals striations on the cube's faces. These striations are consistent with some, but not all, of the symmetries of the cube. The reason for the striations, it turns out, is that the arrangement of atoms inside the crystal is less symmetrical than its external cubic form suggests. Consequently, the pyrite crystal is a cube with texture, or a decorated cube.

One of the more exciting and instructive exercises for older students is to make a cubic kaleidoscope. The cube is divided by its mirror planes into 48 congruent tetrahedra. If a model of one of these tetrahedra is lined with mirrors or some reflecting paper such as mylar, with the triangle belonging to the cube space removed and the opposite vertex snipped off, an entire cube is generated by the reflections. Reflecting mylar pasted onto cardboard or heavy paper will work well; only three of the four tetrahedral walls should be constructed so that you will be able to see inside. Figure 14 shows how to construct such a kaleidoscope.[18]

Using Symmetry

If all we learn about symmetry is to identify it, we miss the whole point. Symmetry is an effect, not a cause.[19] Why are so many natural structures symmetrical? For example, what atomic forces ensure that

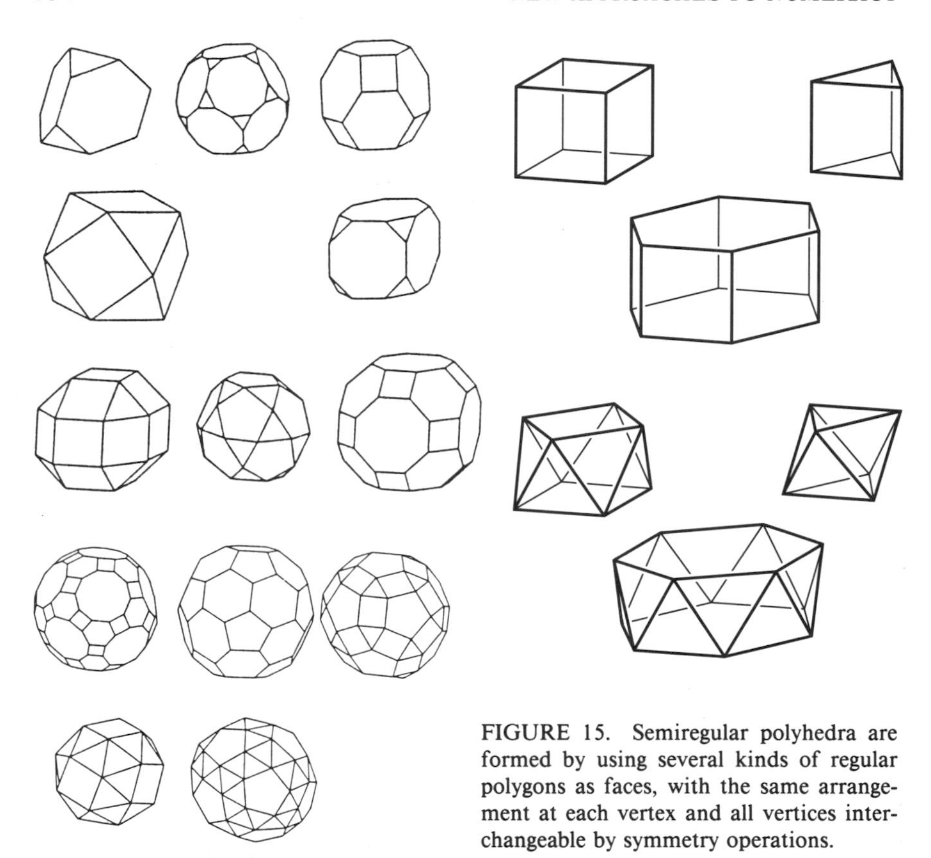

FIGURE 15. Semiregular polyhedra are formed by using several kinds of regular polygons as faces, with the same arrangement at each vertex and all vertices interchangeable by symmetry operations.

the arrangements in crystals will be orderly? Although these are profound and largely unsolved problems, a good working answer was given over thirty years ago by James Watson and Francis Crick in describing their discovery of the structure of DNA:[22]

> Wherever, on the molecular level, a structure of a definite size and shape has to be built up from smaller units ... the packing arrangements are likely to be repeated again and again and hence sub-units are likely to be related by symmetry elements.

In other words, nature builds modular structures that organize themselves according to certain rules. Repetition of the rules tends to lead to arrangements of modules that we call symmetrical.

Polyhedra provide a wealth of excellent examples of arrangements that are repeated again and again. When you build a cube with cardboard squares by attaching three squares to each corner, you are constructing a shape that satisfies a certain packing arrangement: it must be made of congruent regular polygons, and it must have the same number

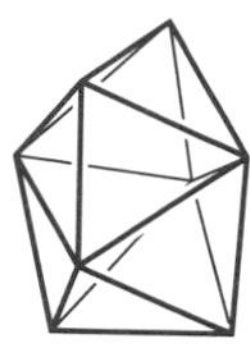

FIGURE 16. Convex deltahedra are formed from equilateral triangles arranged with differing types of vertex arrangements: three, four, or five triangles may be joined at a vertex.

at each corner. By generalizing this construction to other polygons, we obtain the five regular polyhedra (Figure 1). The arrangements can be further generalized to include the semiregular polyhedra (Figure 15), in which more than one kind of regular polygon can be used, and the convex deltahedra (Figure 16), all of whose faces are equilateral triangles but whose vertex arrangements need not all be the same.[17]

The cover design for the biological journal *Virology* contains an icosahedron. The story of the discovery of icosahedral symmetry in viruses and the ongoing efforts of scientists to link that symmetry to their subunit structures is very instructive.[17] Viruses are tiny capsules that contain an infective agent. The capsule is composed of protein subunits that group together to form a closed shell. Watson and Crick realized, in the course of early X-ray investigations into virus structure, that the shells of many viruses had polyhedral or helical forms. Subsequent studies showed that the polyhedra were often icosahedra, and this suggested many attractive models for the arrangement of the protein subunits. But more recently these models have been found to be incorrect. The connection between packing arrangements and overall symmetry in viruses remains an unsolved problem. Problems such as these lead also to new developments in mathematics: they force mathematicians to rethink their definitions and to broaden the scope of their investigations.

Lattices

From earliest times the beautiful shapes that we call crystals have been a source of wonder and admiration. Why do they have polyhedral forms when most other natural structures do not? Quartz crystals were the first to be studied; at first they were thought to be pieces of permanently frozen ice. (It is instructive that our word "crystal" comes from the Greek word κρισταλλος, which means ice.) By the seventeenth century, scientists began to suspect that the shapes of crystals reflected an orderly, patterned, internal structure. Long before the development of modern atomic theory it was suggested that crystals are made of stacks of tiny spheres that represented the basic particles of the structure, whatever those might be. Later the particles were represented as tiny bricks (Figure 17). Sphere packings and bricks (not necessarily rectangular) are still important models for crystal structure.

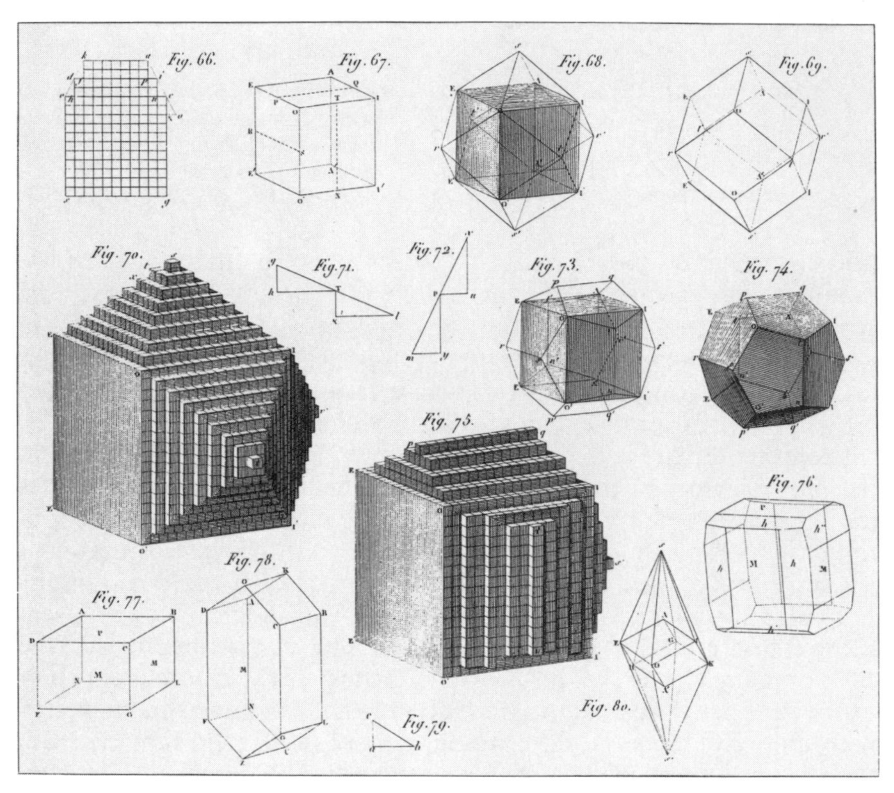

FIGURE 17. An 1822 concept of crystal structure in which various crystal shapes are imagined as being built from tiny rectangular bricks.

Whether we use spheres or bricks, the important idea is that of an orderly array. Let us explore this a little further. A one-dimensional *lattice* is a set of points equally spaced along a line. (Although we can draw only part of the set, we assume that it goes on forever.) All one-dimensional lattices are essentially alike, differing only in the spacing between points. But there are two basic kinds of two-dimensional lattices: one in which the points of the rows lie directly above one another, the other in which they are shifted horizontally (see Figure 18). Each point of a lattice "occupies" a certain portion of the plane, the region nearer to it than to any of the other lattice points. These regions, called Dirichlet domains, display the symmetry of the lattice in a corresponding brick model. The Dirichlet domains in two dimensions—the bricks—are always quadrilaterals or hexagons; within each lattice the regions about each of the points are congruent.

Lattices describe the underlying symmetries of patterns. Draw a one-dimensional lattice on two or three different sheets of tracing paper, and

FIGURE 18. The symmetry of two-dimensional lattices is displayed by their Dirichlet domains, polygons centered at each lattice point which enclose the region of the plane closer to the enclosed lattice point than to any other. These polygons may be quadrilateral or hexagonal; for a given lattice they are all congruent.

use them to create two-dimensional lattices. You will quickly discover that you can change the symmetry of the lattice by shifting the relative positions of the rows: you can check the symmetry by recalculating the Dirichlet domains. No matter what you do, the symmetry will always be of one of the five types shown in Figure 18. It is an important fact that every two-dimensional repeating pattern, whether it is an arrangement of points or ellipses or polygons, a wallpaper pattern, or an Escher-like tiling of the plane, can be interpreted as a decoration of the Dirichlet domains associated with a lattice that belongs to one of these five symmetry types.

This simple observation raises a wealth of interesting questions. What kind of packing arrangements can we create if we replace the points by other shapes? What shapes can be fitted together without gaps to form orderly patterns? What do we mean by orderly? What are the possible ways to extend arrays to three dimensions? It turns out that there are only a small number of solutions to problems such as these, which explains why the same patterns reappear so often in crystal structures, trusses, biological tissues, honeycombs, wallpaper, textiles, and tiled floors.

Three-dimensional lattices have been used by mathematicians and scientists, beginning in the nineteenth century, to try to explain the arrangements of atoms in crystals. In three dimensions there are 14 symmetry types of lattices and 5 combinatorial types of Dirichlet domains (see Figure 19).

It is difficult to overestimate the importance of play with cubes and other blocks. Even one year olds enjoy building taller and taller towers and watching them fall down. Later, children use blocks to build

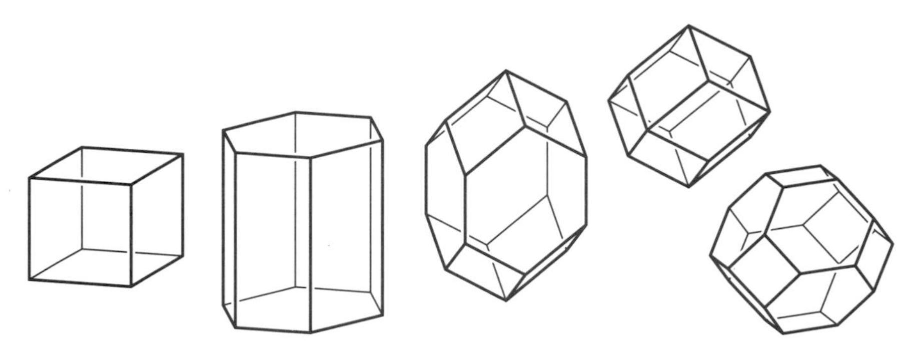

FIGURE 19. There are five combinatorial types of Dirichlet domains for three-dimensional lattices. These five shapes are much less well known than the Platonic solids but are at least as important!

houses, courtyards, and other structures. Young children are likely to have trouble building octahedra, but they can use small cubes to build larger cubes. The smallest composite cube is made up of 8 smaller ones; the next larger is made up of 27; by guessing how this series continues the child gains some understanding of volume. Older children—of any age—can also learn a lot from playing with cubes.

Cubes are the prototypical three-dimensional tile, and many structures, both mathematical and real, are based on it. It is worthwhile to try to build polyhedra out of cubes. For example, try building a regular octahedron by sticking sugar cubes together with glue. The larger you make your sugar octahedron (if it isn't too messy), the closer the stepped faces approximate smooth ones. Building polyhedra from cubes is thus a sophisticated lesson in volume measurement. It is instructive that H.S.M. Coxeter, in his classic work *Regular Polytopes,*[5] refers to the cube of any dimension as the "measure polytope." (The word "polytope" refers to the higher-dimensional analogues of polygons and polyhedra.)

Dissection

An important problem in many fields is how to divide a region into compartments of various shapes. An architect or designer partitions the interior of a building into rooms to serve certain purposes. We all fret over the most efficient way to pack a suitcase or the trunk of a car. A complex living object, such as a plant or a human being, has grown from a single cell that, in the early stages of growth, divided into "daughter" cells that grew and divided again. The study of how dividing cells organize themselves into tissues and then into organs is one of the most exciting frontiers of biology. Some of the issues relate to the geometry of dissection, compartmentalization, and subdivision.

There are many interesting mathematical problems dealing with dissection. One of the most famous theorems in this field says that any polygon can be divided into a finite number of pieces and reassembled to form a congruent copy of any other polygon of the same area. Elementary school children enjoy the challenge of creating shapes with tangrams or other polygonal tiles; imagine the many challenging problems and puzzles that could be devised for older children related to this dissection theorem. More advanced students can discover that the analogous theorem for polyhedra is false; this is another fascinating and important result.

Another intriguing dissection problem is the creation of "rep-tiles," tiles that can be fitted together to form replicas of themselves (Figure 20). Alternatively, we can create such tiles by subdividing one into smaller congruent copies of itself. To create a tiling by rep-tiles, think of the daughter tiles growing to the size of the original one and then subdividing again. Repeating this process over and over again, we create a tiling that is self-similar in a certain sense; many of these tilings have no lattice structure. Is the tiling of Figure 20 lattice or nonlattice?

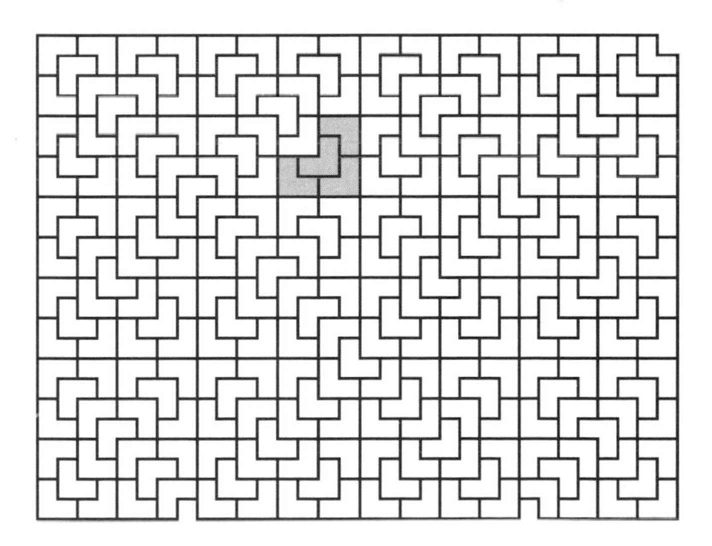

FIGURE 20. "Rep-tiles" are tiles that can be fitted together to form replicas of themselves. They build tilings that are self-similar and that, like this one, may have no lattice structure. Such tilings are of great interest today because they share many strange properties with some newly discovered crystalline materials.

(This is not easy to answer!) Tilings without lattices are of great interest today among mathematicians and solid-state scientists because they share many strange properties with some recently discovered crystalline materials called quasicrystals.

Combinatorial Tools

Combinatorial properties of patterns are also very important because they provide clues to what is possible and what is not. For example, suppose we want to build a tetrahedron—that is, a polyhedron with four faces. How should we start? Before cutting out polygons and trying to tape them together, let's reason out the possibilities. In the first place, all the polygons will have to be triangles, because as we build we will start with one polygon and attach another to each edge. If our first polygon had more than three edges, we would run out of polygons, since we only have four. So to build a tetrahedron we attach a triangle to each edge of our first triangle (Figure 2a), and then to make a closed polyhedron we must fold up the configuration so that the other triangles meet in a point. This means that the edges of the polyhedron must form a network of four triangles.

We can go on from here to discuss properties (e.g., congruence) these triangles might have, but it is important to note that we have already made an important discovery: *every tetrahedron is a combinatorial network of four triangles.* Similar reasoning shows that there are two combinatorial types of pentahedra, polyhedra with five faces (you can find them in Figure 6). There are exactly seven types of hexahedra (Figure 21), including, of course, the cube. It is a challenge for students to discover why there are no more.

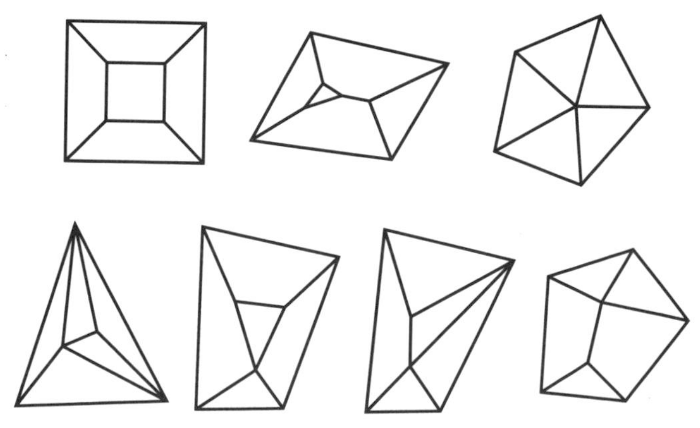

FIGURE 21. There are seven combinatorial types of hexahedra. Try to visualize them as three-dimensional polyhedra!

The combinatorial properties of shapes are sometimes more fundamental than their metric properties. If we try to build a convex polyhedron out of hexagons, we will never succeed: such polyhedra are combinatorially impossible. It's better to know this in advance! A few years ago a "World Sports Day" poster featured a giant soccer ball that appeared to be made entirely of hexagons. The designer did not realize that she had drawn an impossible figure!

The fundamental theorem of combinatorial theory for polyhedra is Euler's Theorem, which is valid for every convex polyhedron (and some others): the sum of the number of faces plus the number of vertices is equal to the number of edges plus two. This can be written succinctly as

$$F + V = E + 2$$

where F is the number of faces (or cells) of the network, V is the number of vertices, and E is the number of edges. (It is easy to verify this equation with the networks in Figures 1, 6, 19, and 27). Euler's Theorem is easy to discover (with guidance), easy to teach, and, for more advanced students, not difficult to use. The theorem and its many corollaries and generalizations are important tools for enumerating the combinatorial properties of objects.

REPRESENTATION

A third important tool in the study of shape is *representation.* In everyday life as well as in science, mathematics, and art, we deal not only with shapes themselves but also with many kinds of representations of shapes—models, photographs, drawings. The tools of representation include the ability to understand scale models; to read maps; to understand shadows, sections, and projections; to reconstruct shapes from their images; to draw accurately; and to use computer graphics. The underlying issue is the same in each case—to determine the relation between a shape and its image or between different images of the same shape.

Models

The simplest representation of a shape is a model of it, built to an appropriate scale. A spherical globe is a model of the earth, or of the moon, or of any planet. A globe is not an exact replica of the earth, but an approximate one that displays certain features of the earth quite well. It is approximate because it is perfectly round, which the earth is not. Besides, it is constructed on such a small scale that even our largest cities appear as tiny dots. But every child growing up in this

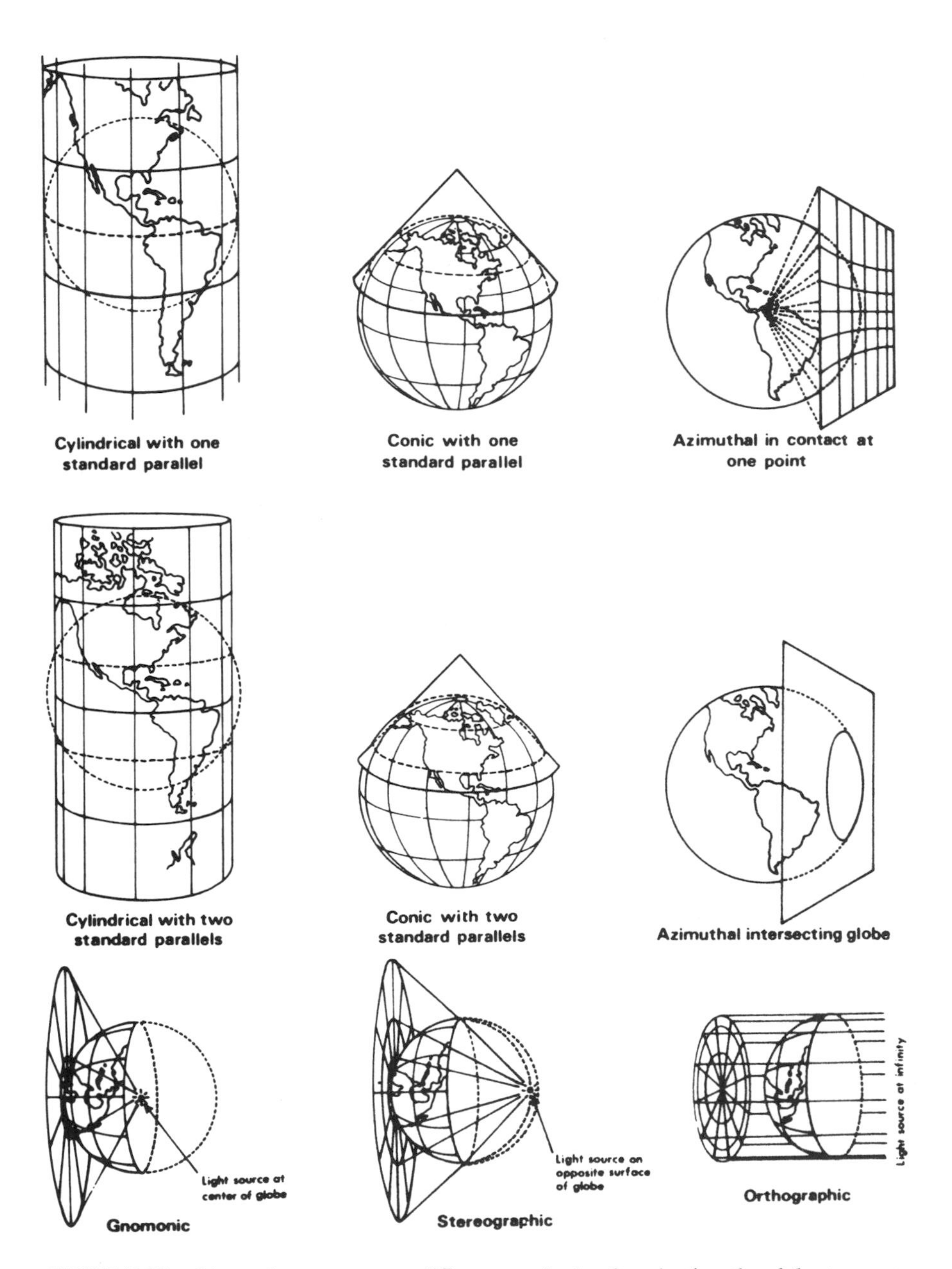

FIGURE 22. Mapmakers use many different methods of projecting the globe to create flat maps. The choice of projection determines many of the map's features.

space age understands that the globe is a model of the earth. Most models are approximate in the sense that they ignore some details in order to present key features more vividly. Making a model entails making a selection of which features are to be emphasized; this point merits classroom discussion.

Maps

Interesting questions about the relation between shape and image arise, for example, in the study of maps. Why do we use both globes and flat maps? The answer is simple: they are useful for different purposes. Although a globe and a flat map represent the same thing, namely the earth, they display its properties very differently.

Flat maps can represent small regions of the earth quite well, since part of the surface of a spherical object can be closely approximated by a plane. But the representation gets worse and worse as you try to increase the area represented by the map. The relation between globes and flat maps leads quickly to very fundamental geometrical questions. You can't make a sphere by folding up a sheet of paper, so to make a flat map, you have to project the globe in some way. Mapmakers use several different projection methods (Figure 22). Thus a map of the earth is an approximation of a spherical surface, an approximation that gets worse and worse as the portion of the globe being mapped is increased. Every flat map necessarily distorts angles, areas, or both. Like a mathematician considering more general kinds of maps, every mapmaker must compromise by deciding which features of representation are most important for particular purposes.

Shadows and Lenses

Shadows, perhaps the most familiar examples of images, are nevertheless rather subtle because they distort contour as well as size. The interesting question is to determine what sorts of distortions can occur, and why.

Young children can learn a great deal by observing their own shadows. To create a shadow, you need a light source (if you are outdoors, it is the sun), an object (you), and a screen (the ground or a wall). The shadow is your projection onto the screen, and what that projection looks like depends on the positions of the light, the object, and the screen. Older students can experiment, varying the positions of the light, the screen, and the object that blocks the light to produce the shadow. From this they can discover which properties of shapes are preserved and which

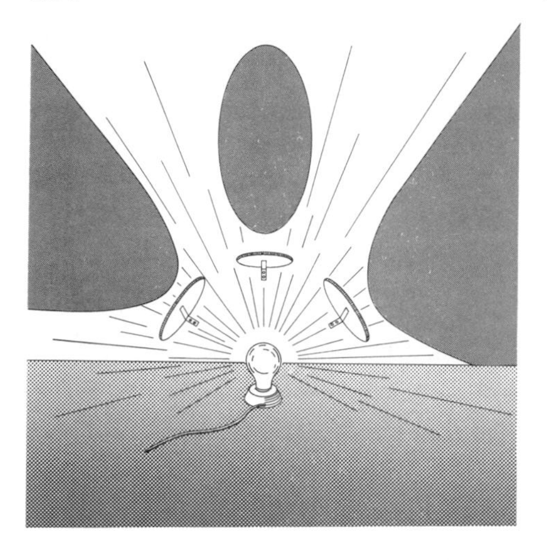

FIGURE 23. The circle, the ellipse, and the parabola as shadows cast by a circle on screens in different positions.

are lost under this kind of projection. For example, all the conic sections can be created as projected shadows of a circle (Figure 23).

At a more advanced level we can think of shadows as maps in which only the outline of the map is retained. From this perspective the principal difference between a map of the earth and your shadow on the wall is the object being mapped.

Lenses, too, distort shape but in more predictable ways. Eyeglasses, slide projectors, telescopes, microscopes, and cameras are only a few of the tools through which lenses enter our lives. Indeed, lenses in our eyes provide our only access to visual images. The study of lenses involves many principles of geometry that can be taught at every level from kindergarten through high school and beyond.

Drawing

In every culture and in every era artists have grappled with the problem of representing three-dimensional shapes on two-dimensional surfaces. The solutions they have found are, in many cases, the same as those of the mapmaker. For example, in Figure 24 the artist is literally making a map of the shape he sees before him. The device he is using is easy to make and can be used in the classroom with good results. Perspective drawing is another example of mapping.

Before the camera became available to everyone, drawing was widely taught. Today very few people know how to draw accurately, and, consequently, they no longer notice things as carefully as they once did. A few years ago there was great embarrassment (or should have been) when Branko Grünbaum discovered that the icosahedral logo of the

FIGURE 24. Albrecht Durer's sketch of an artist making a map of the shape he sees before him.

Mathematical Association of America, which appeared on all of its publications, was inaccurately drawn (Figure 25); this error had escaped detection for several years even though it was seen regularly by thousands of mathematicians. If visual illiteracy is so widespread even among professional mathematicians, future generations run the risk of really believing that they live in an Escher-like impossible world (Figure 26)!

In his article on the misdrawn icosahedron Grünbaum[11] presented a small sample of his collection of badly drawn textbook figures (see Figure 27). Looking at them with a trained eye is enough to make one laugh or cringe (or both). But how many of us could do better? Indeed, how many teachers of mathematics can even draw a respectable cube? These gaffes presumably would not have occurred were authors and graphic artists more familiar with the principles and practice of perspective and projection. For many years technical drawing has been relegated to courses in the fine and industrial arts when, in fact, they are essential for all students.

FIGURE 25. The icosahedral logos of the Mathematical Association of America: the old one is badly drawn; the new one is accurate. The error escaped detection for many years. Can you tell which is which? (Hint: In drawing a projection of a three-dimensional figure on a plane, parallel lines should stay parallel or else intersect in a single point.)

FIGURE 26. *Belvedere,* by M.C. Escher, provides a visual commentary on the subtleties of representing a three-dimensional scene on a two-dimensional piece of paper.

Image Reconstruction

If the artist's problem is to represent a three-dimensional shape on a flat surface, the viewer's problem is to recognize what shape the image is supposed to represent. A visit to an art gallery is an exercise in image reconstruction. So is the physician's task of reading an X-ray or a space scientist's task of interpreting photographs of the surface of Mars.

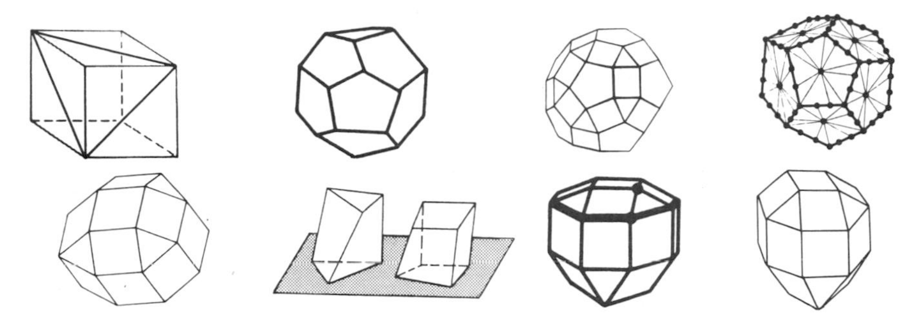

FIGURE 27. Some badly drawn figures, all taken from published books dealing with geometry and related subjects.

The painting, the X-ray, and the photograph are maps of shapes, which we need to be able to read "in reverse." This subject, closely related to the problem of visualization, is of great importance in the study of shape, but it has not been organized in a way that can be used in school. Here is a challenge: to bring to students of mathematics the wealth of material related to shadows, cross sections, and projections. In addition to ideas now taught in art classes and in the industrial arts, students could learn about criteria for deciding whether a projection is properly drawn (i.e., whether a diagram is in fact a projection of a three-dimensional form). They could learn the principles of the stereoscope and why stereoscopic pairs appear three-dimensional to us. They could also learn to deduce symmetry and topological properties of a three-dimensional shape from its two-dimensional representation. A discussion of optical illusions and "impossible" figures can lead to many important insights. Those familiar with the combinatorial properties of polyhedra can study their representation through planar graphs and can try their skill at reconstructing corresponding three-dimensional forms.

Computer Graphics

The computer is not a substitute for real three-dimensional models. Images on the computer screen, even the so-called 3-D images, are meaningful only if the viewer has extensive prior experience with three-dimensional structures. On the other hand, computer graphics can be fascinating to students and can generate strong interest in the study of shape. Good software can thus be invaluable in the study of shape and should be used when appropriate. Moreover, every person should know something of the geometry that underlies computer graphics—above all, coordinate geometry—in order to use graphics packages intelligently and critically.

In summary, the creation of images and the reconstruction of shapes from their images is central to the study of shape. All of the many facets of representation can be organized under the concept of mappings. "Real" maps are only one example; shadows, sections, images seen through lenses, images produced by projection, images produced by reflection, and images rendered by the graphic or photographic artist are others. As increasingly detailed images of the very large, the very small, and the formerly hidden are made visible by modern technology, the need to understand this broadened concept of mapping becomes increasingly urgent.

Mapping is a major theme of contemporary mathematics because it provides a useful and illuminating way to organize relations among shapes and patterns (including very abstract ones). It also helps us to make our classification systems precise. Congruence and similarity can be described in the language of maps. For example, the shapes in Figure 3a can be transformed into one another by a mapping that preserves their combinatorial structure; the shapes in Figure 3b are related because they have the same set of symmetries (which are mappings of the objects on to themselves); the shapes of Figure 3c can be transformed into one another by a mapping that is a continuous deformation.

VISUALIZATION

Visualization is a broad subject with implications for many aspects of our lives. It is centrally important to all of mathematics and has been so throughout history. Mathematics made a great advance with the invention of numerals, which are visual representations of numbers. Certainly one of the major mathematical achievements of the last several hundred years was the development of analytic geometry, which enabled us to combine visual and formal mathematical thought.

Obviously, visualization is very important in the study of shape. But it is also important for all of mathematics. To study change, we need to see it; to study data, we examine various graphical representations. We try to grasp the concept of higher dimension by drawing pictures and by making models. Even the properties of numbers can be illuminated by visual representation—that is what the number line is for. But it is not true that we instinctively know how to "see" any more than we instinctively know how to swim. Visualization is a tool that must be cultivated for careful and intelligent use.

It may be helpful to retell a very old story about Galileo's discovery of mountains and craters on the moon, a discovery that helped to change forever the way we view the universe and our place in it. "Following Aristotle, Europeans of the Middle Ages and the Renaissance believed that the moon was a perfect sphere, the prototypical shape not only of the visible planets and stars but of the entire universe," explains the art historian Samuel Edgerton.[6]

The problem, thus, was not to determine its shape, which all accepted, but to explain the mottled appearance of its surface, that "strange spottednesse," as Harriot called it. Some ancient authorities had explained the spots by arguing that the lunar surface was like a gigantic mirror reflecting the lands and seas of the earth. Others had claimed that the moon was composed of transparent substance with some internal denser matters giving off varying amounts of light.

Galileo found another explanation:[8]

I have been led to the opinion and conviction that the surface of the moon is not smooth, uniform, and precisely spherical as a great number of philosophers believe it (and the other heavenly bodies) to be, but is uneven, rough, and full of cavities and prominences, being not unlike the face of the earth, relieved by chains of mountains and deep valleys.

Thomas Harriot was an English astronomer who had also been looking through a telescope at the moon at the same time that Galileo made his discoveries. Harriot's sketches show, however, that the "strange spottednesse" did *not* look like mountains and valleys to him (Figure 28).

How could it happen that Harriot and Galileo, looking at the same object through comparable telescopes, did not "see" the same thing? True, Galileo was the greater genius, but this fact alone is not very illuminating. Edgerton suggests a more persuasive reason: Galileo was a trained artist, skilled in the use of perspective and chiaroscuro, the rendering of light and shadow. Thus "Galileo did indeed have the right theoretical framework for solving the riddle of the moon's 'strange spottednesse.' Unlike Harriot, he brought to his telescope a special 'beholder's share' (as E.H. Gombrich would say); that is, an eyesight educated to 'see' the unsmooth sphere of the moon illuminated by the sun's raking light."

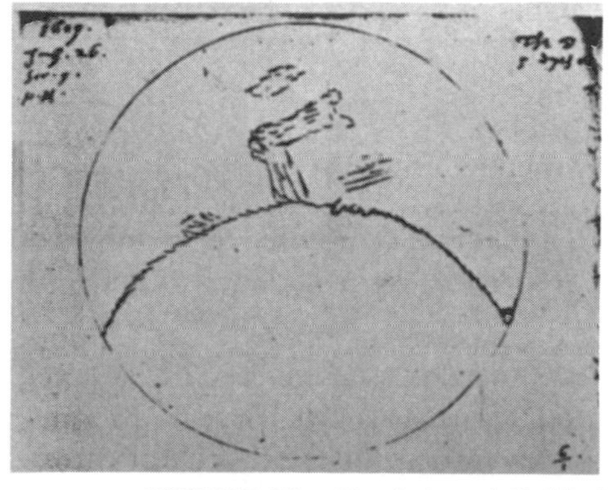

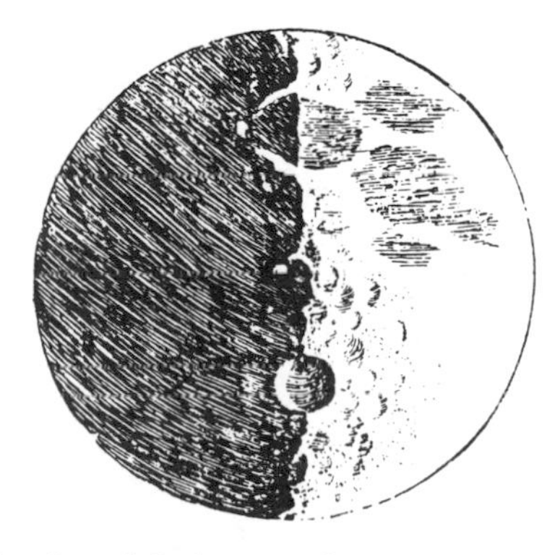

FIGURE 28. Harriot's and Galileo's sketches of the lunar surface.

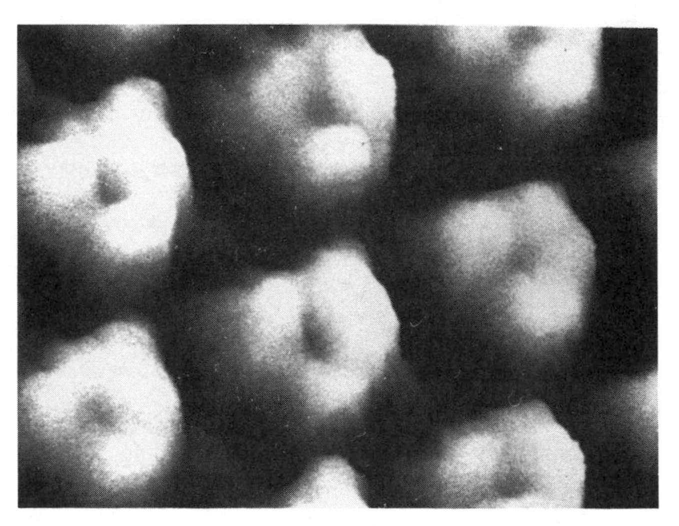

FIGURE 29. The benzene ring was made visible on the atomic scale for the first time in 1988. This image was produced by a scanning tunnelling microscope. An educated eye can see the triangular patterns of electrons connecting pairs of the six carbon atoms in each ring.

"Galileo's telescopic discoveries opened the eyes of Europeans everywhere," continues Edgerton. And, as his notebooks show, "even Harriot 'saw' shaded craters once he was aware of the Florentine's observations."

Today, we all see mountains and valleys when we look at the moon. But would we see them if we didn't already know what we were supposed to see? And what do we "see" when we look at the images presented to us by modern technology? The educated "beholder's share" is just as essential today as it was in Galileo's time. "Whether the object is a virus seen through an electron microscope, a distant galaxy explored by radio telescope, or a fetus observed in the womb by means of ultrasound, theoretical assumptions have to be made before the raw data can be translated into an image," writes Hans Christian von Baeyer in a recent issue of *The Sciences*.[2]

Von Baeyer goes on to point out that this translation must be done by the educated eye as well as by the internal workings of the computer. A case in point is the first atomic-scale image of the "hexagonal" benzene rings, which was produced for the first time in 1988 (Figure 29). Can you see the hexagons? Or do you see some lumpy donuts? Or spherical triangles? Scientists were able to find the triangular traces of the hexagonal structure because they already knew that they were there.

Rudolf Arnheim stressed the importance of visualization in science in his aptly titled *Visual Thinking*:[1]

The lack of visual training in the sciences and technology on the one hand and the artist's neglect of, or even contempt for, the beautiful and vital task of making the world of facts visible to the enquiring mind, strikes me, by the way, as a much more serious ailment of our civilization than the "cultural divide" to which C.P. Snow drew so much public attention some time ago. He complained that scientists do not read good literature and writers know nothing about science. Perhaps this is so, but the complaint is superficial .... Snow's suggestion that "the clashing point" of science and art "ought to produce creative chances" seems to ignore the fundamental kinship of the two.

Like the weather, everyone talks about visualization, but no one does much about it. Visualization is not a simple matter: it is a deep subject, properly the domain of physiology and psychology and still not well understood. Nonetheless, it is easy to teach shape as an important first step in developing powers of visualization. The simplest way to teach students to visualize is to provide them with a rich background of hands-on experience with shapes of many kinds. A serious study of image reconstruction would also be a step in the right direction.

CURRICULAR ISSUES

Students should learn to recognize the patterns of shape, to understand the principles that govern their construction, and to be able to move easily back and forth between shapes and their images. Although the study of shape seems to fall between the cracks of traditional subjects, the new *Curriculum and Evaluation Standards for School Mathematics*[15] of the National Council of Teachers of Mathematics reflect an emerging consensus that this situation must be improved.

The study of shape must be more than the sum of its parts; an integrated view of shape can help accentuate the whole subject. One possible approach is illustrated by the chart in Figure 30.

Forging Connections

Rethinking the subject as a whole provides us with an opportunity to forge substantive connections between the study of shape and the role of shape in the real world. We can take seriously Arnheim's plea for integrating art and science. We can also reduce the mystery of some of our contemporary technology. The principles of the electron microscope, the radio telescope, and ultrasound are not wholly beyond the scope of the K–12 curriculum; high school students can, if we wish, learn the foundation necessary to understand the action of these and other modern imaging techniques.

Indeed a focus on shape makes many aspects of modern technology much more accessible than is commonly supposed. Here are just three

A Structure for Shapes

Identification and Classification

ELEMENTARY:	INTERMEDIATE:	ADVANCED:
Circles	Spheres	Surfaces
Plane Polygons	Zig-zag and star polygons; knots	Helices, spirals, cylinders, tori, Mobius bands
Polyhedra	Polyhedra	Polyhedra
Puzzles	Tiling the plane with polygons	Escher-like tilings
	Networks	Simple crystal structures
Congruence; similarity		
Soap bubbles	Soap bubble clusters	Orientation, genus texture

Analysis

ELEMENTARY:	INTERMEDIATE:	ADVANCED:
Mirror symmetry; rotational symmetry	Two-mirror kaleidoscopes	Polyhedral kaleidoscopes
Congruence	Symmetry of finite figures	Symmetry as an organizing principle; transformation geometry
Paper-folding; patterns	Dissection; puzzles	
Similarity	Rep-tiles; fractals	Exploring fractals
	Natural patterns	Scale in biology
Constructing and deconstructing polyhedra	Regular and semiregular polyhedra	Euler's formula for polyhedra
Linear/volume measurement	Angle measurement	Fundamentals of plane and 3-D geometry
Making quilts and mosaics	Tiling the plane with polygons	Lattices; elementary tiling theory

Representation and Visualization

ELEMENTARY:	INTERMEDIATE:	ADVANCED:
Model-making	Model-making	Model-making
Drawing, reading, and using simple maps	Relief maps and level curves	Cross-sections of 3-D shapes structures
	The globe	Geometry of the sphere; projections; maps
Shadows	Shadow geometry	Images and image reconstruction; impossible figures
Drawing	Perspective drawing	Technical drawing; stereoscopes
Scale projectors	Telescope and microscope	Lens geometry; the camera
	Plane coordinates	3-D coordinates
Turtle geometry	Exploring geometry with the computer	More computer graphics

FIGURE 30. An arrangement of topics related to shape that provides structure and coherence to what might otherwise appear as an arbitrary collection of quite disparate topics.

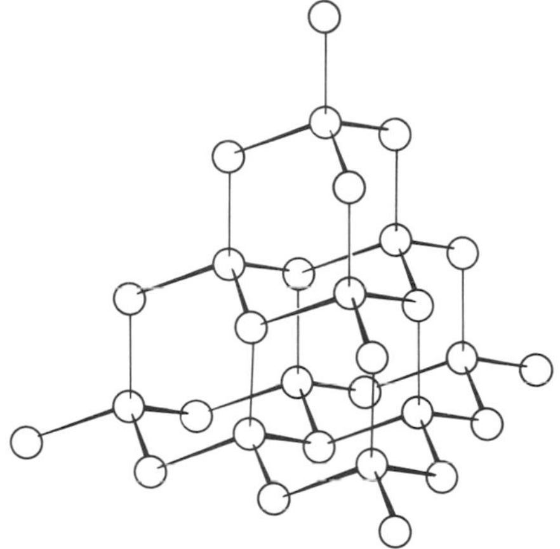

FIGURE 31. The structure of crystalline silicon. It is made entirely of zig-zag hexagons. This is also the structure of diamond (with carbon atoms at these positions instead of silicon).

examples of important shapes whose key features could easily be taught in our schools.

The *silicon chip,* which has transformed the industrialized world in just a few decades, is based on a structure that is a carrier of incredibly miniaturized circuits. Although the circuits themselves are complex, the crystal structure of the silicon that houses them is a simple modular structure.

For example, crystalline silicon is built of linked zig-zag hexagonal rings (Figure 31), which are easy to make and instructive to study. In the silicon structure the rings are linked to form cage-like polyhedra. Elementary school children can learn to build and identify these substructures, middle school children can learn to put them together, and high school children can study the relation between the silicon structure and the properties that make it so useful.

The *CAT scan* and other forms of computer-assisted image reconstruction have revolutionized medical diagnosis in recent years. While diagnosis by X-ray is an exercise in reading shadows, diagnosis by CAT scan is an exercise in reconstructing images from their cross sections. Like the circuitry on a silicon chip, the image reconstruction used in this technology is a complex process, but the simplest geometrical principles that underlie it are easily understood.

Here again we find that the same geometric principles are central to many fields. For example, the construction of shapes from sections and shadows has been the task of architects and builders for centuries. While it is not feasible to bring a CAT scan machine or a construction site into the classroom, many projects suitable for school can help students understand the relation between shadow or cross section and shape.

Snowflakes, especially the feathery ones, are enchanting. Children often learn to make paper snowflakes in school, an exercise that can easily

FIGURE 32. Branched snowflakes reveal the familiar hexagonal symmetry of ice crystals repeated fractal-like at every scale.

be extended to a study of their symmetry. The hexagonal symmetry of the snowflake provides an introduction to the symmetry of polygons; it is an ideal subject for the elementary classroom.

But the snowflake has much more to teach us. In the first place a snowflake looks like a pattern we might see in a kaleidoscope, and so it is. This suggests a study of the kaleidoscope, which, as we have seen, is an application of the principles of mirror geometry. These same reflection principles undergird contemporary technology: one need only think of the reflection beams of burglar alarms and lasers or of radar and sonar. Middle school children can easily understand and appreciate such applications. At the high school level the emergence of hexagonal symmetry from aggregates of water molecules can be explored and so can the crystals' dendritic growth, or branching.

The branching of the snowflake is as characteristic as its symmetry and is equally significant in the study of shape. First, corners of the snowflake sprout beyond a hexagonal "core." Then these branches themselves sprout branches, the branches of the branches branch, and so forth (Figure 32). The result is a structure in which a certain feature—branching—is increasingly repeated on a smaller and smaller scale. If this process could be repeated indefinitely, the result would be a self-similar structure; indeed, the snowflake is a fractal at an early stage of its development.

Geometry

The role of geometry is a perennial issue in mathematics education at all levels from elementary school to graduate school. For many years geometry has been the problem child of the mathematics curriculum. A glance through the National Council of Teachers of Mathematics' 1987 Yearbook *Learning and Teaching Geometry*[14] suggests some of the many questions involved. The problem with geometry is due in part to lack of agreement on what geometry is and why we should study it. Do we study it to learn disciplined thought? To prepare students for other subjects? Or is it because there is important content in the subject itself?

Most high school geometry texts do point out examples of geometric forms in nature, science, technology, and art, although none of these connections is ever explored in any depth. The synthesis of method and content is usually unsuccessful. Our geometry courses are uneasy compromises between many important but very different goals: teaching deductive reasoning, proving theorems of Euclid, introducing problem solving, teaching visualization, and preparing the students for calculus. The continuing debate indicates that none of these goals is particularly well served in the present situation.

In almost all of these debates the teaching of geometry is defended on the grounds that it serves external purposes, rather than on the importance of the subject in its own right. For example, a recent article on similarity[7] justifies the teaching of similarity with the following rationale:

> Similarity ideas are included in many parts of the school curriculum. Some models for rational number concepts are based on similarity; thus, part of the students' difficulties with rationals may stem from problems with similarity ideas. Ratio and proportion are part of the school curriculum from at least the seventh grade on, and they present many difficulties to the student. Standardized tests include many proportion word problems. Verbal analogies (a:b::c:d) form parts of many intelligence tests. Similar geometric shapes would seem to provide a helpful mental image for other types of proportion analogy situations.

All of these reasons are valid ones, but there is a striking omission: the principal reason for teaching similarity ought surely to be that it is of profound importance in understanding shape.

Meanwhile, outside the halls of education, the computer revolution is rapidly changing the world in which we live. These changes are placing new demands on the curriculum, demands that are just beginning to be heard in the schools. The revolution in the study of shape and form made possible by the computer suggests that what we need is not just a better compromise for geometry, but a new and coherent mathematics curriculum that integrates shape into the entire course of study.

Is Euclid to stay or go? This is not a useful question. We need to ask instead what we want our students to know and why. Euclid realized that careful reasoning about shape requires careful statements of definitions and assumptions and very careful argument. In order to analyze shapes, students must know how to measure lengths, areas, and volumes as well as planar and dihedral angles. They need to know properties of parallel and perpendicular lines, basic angle theorems, and fundamental properties of figures. Moreover, they need to know how, with ruler and compass, to construct such standard figures as equilateral triangles, regular hexagons, and squares.

The study of shape therefore overlaps the traditional geometry curriculum, but it cannot be subsumed under it as a brief module or extracurricular activity. The shapes that students need to understand today, and the things that they need to be able to do with them, are too vast a subject for that. Moreover, there is considerable difference of emphasis and purpose.

Traditional geometry shares more than just historical roots with classical civilization. Its role in school in some respects is analogous to the curricular issues of classical versus modern languages. A student who studies Latin or Greek learns rigorous thought and some important history and also acquires the basis for many modern languages, including our own. Modern, spoken languages, on the other hand, are less rigorous yet more fluid. They are the living flexible languages that people actually use in everyday life. Ideally, students should learn both classical and modern languages, although few have the time or opportunity for both. A watered-down Latin course enlivened with examples of cognate words in Italian, Spanish, or French is not the solution to the problem.

All of the virtues of Latin and Greek are shared by classical Euclidean geometry. For over 2000 years Euclid's *Elements* has served not only as the cornerstone of geometry but also as the very model of mathematical reasoning. Deductive reasoning from axioms has been very fruitful, not only for mathematics but also for science and philosophy. For example, it was questions raised by Euclid's axioms—rather than observation of the real world—that led to the discovery of non-Euclidean geometry, which subsequently became the central tool in studying the large-scale structure of the universe. Classical geometry has not lost its value, but other needs require that we also introduce the mathematical counterpart of modern language courses into our curriculum.

Studying Shape

Shape is a subject that cuts across many parts of mathematics and science. It offers a rich variety of possibilities for imaginative, exploratory instruction—from building models to using computers, from observation to experiment, from manipulation to calculation. Shape holds extraordinary potential for enhancing the quality of mathematics instruction, in several different ways.

The study of shape is interdisciplinary. As we have already noted, many subjects in which shape plays a role are not usually thought of as mathematics in a narrow or restricted sense. For example, problems of size and scale do not belong exclusively to mathematics. They lead to all sorts of questions that send us to the library or to colleagues in other departments. Could there ever have been giants? Could there be people as small as mice? Our myths show that these questions are older than our recorded history. The answers are not straightforward applications of similarity. A giant could not be supported by his legs if they were exactly similar to our legs; instead, the bone mass has to be increased disproportionately. This complication makes the study of biological scale more fascinating than it would be if the answers were simple.

Perspective is taught in art classes; geometrical optics is a branch of physics; similarity and other transformations are central concepts of biology; chemists build polygons and polyhedra to model the structure of molecules. Even within mathematics, shape is interdisciplinary: it requires visual and computational skills, logical thought, and many other tools. Teaching shape in a coherent, meaningful way can stimulate close cooperation among teachers of many subjects.

The study of shape suggests projects cutting across several subjects. For example, the study of similarity can be nicely complemented by a study of lenses, requiring an excursion into physics. Even the names associated with many of the laws of geometric optics (e.g., Fermat's principle) stand as testimony to the fact that today's disciplinary borders have not always been so high. It is precisely because we find the same shapes everywhere that we need to study them—as part of mathematics in many different contexts.

The study of shape is a laboratory subject. All of us, children and adults, learn about shapes by making them and studying models (Figure 33). As an ancient proverb says, "I hear and I forget; I see and I remember; I do and I understand."

If we wish to build a shape—a cube, a scale-model house, or a spiky star polyhedron—we have to be able to cut out and assemble pieces of the correct sizes. This is one of the reasons that basic geometry (angle measurement, parallel lines, and so forth) remains indispensable.

FIGURE 33. The eminent geometer H.S.M. Coxeter studying a model. Coxeter has devoted his life to discovering patterns in shapes.

Building models, in this very concrete sense, is one of the best ways to unify theory and practice.

Hands-on experimentation is essential. For example, when we make a cube with our own hands, we gain much more insight into its metric, combinatorial, and stability properties than if we just look at one. If instead of cardboard squares we make the cube from plastic straws stuck in balls of putty or in marshmallows, the cube will wobble. Though less elegant, the wobbly cube is not a "bad" model. On the contrary, it is a useful one because it teaches something about rigidity and flexibility. It also teaches something about the shapes into which the cube can be transformed while maintaining its combinatorial structure.

Everyone seems to agree that models and "manipulatives" are valuable tools in the classroom. But too often one hears the lament that "if only models were introduced early enough, we wouldn't have to use them later on." This unfortunate attitude masks two implicit but very inaccurate assumptions. First, that gross morphological shape is the main thing that we learn from models and that it can all be learned in elementary school: if you've seen one cube (once), you've seen them all. This, of course, is nonsense: the humble cube plays a key role in the study of volume, congruence, symmetry, and modular structures.

The second assumption is that the main purpose in studying a model is to develop our powers of abstract reasoning; here the model plays the role of training wheels on a bicycle. Certainly we want our students to understand the sense in which a particular cube represents the general concept of a cube. But even once this is understood, most of us still have a lot to learn from real models.

Ideally, shape should be taught in a laboratory setting. At the very least, every school should have a laboratory where students can explore shape. A shape laboratory should include work tables, drawing and construction equipment, three-dimensional models of many kinds, materials for building them, and places to display them. If possible, it should include computers with graphics capabilities. Textbooks should be supplemented with workbooks, project material, and interactive computer graphics programs.

The study of shape is for everyone. It is often said that studying shape is ideal for slow learners. Certainly it is true that students who have trouble with axioms and abstractions will find a hands-on, problem-oriented shape curriculum less difficult and more meaningful. The misconception lies on the other side of the coin—the widespread belief that more advanced students do not need to study shape.

We do not have to look further than today's newspaper for evidence of the folly of this belief. "Supercomputer Pictures Solve the Once Insoluble," proclaimed the headline of a recent article on the front page of the *New York Times.*[13]

> Scientists who are using the new supercomputer graphics say that by viewing images instead of numbers, a fundamental change in the way researchers think and work is occurring. "The human brain is the best pattern recognizer in history," says Heinz-Karl Winkler, a Los Alamos National Laboratory physicist. "We can use it to visually scan vast quantities of data. We can zero in on a structure in an image and distinguish between important things and unimportant things."

It is our best students, not our weakest ones, who will be using supercomputers to study the shape of data and scientific images. How will they know how to distinguish important from unimportant things in a structure if they have never studied structure at all?

The study of shape is fun. Students enjoy working with shape, as we all do. In teaching shape, especially in a workshop setting, a teacher is unlikely to encounter the lack of motivation or the resistance that sometimes arise in geometry courses. Unfortunately, fun is suspect in some educational circles. One effective way to answer questions about the educational value of exploring shape is to hold an open house in the shape laboratory so that doubters can become converts by getting involved with the material themselves.

The study of shape is open ended. In a time of rapid change the study of shape facilitates open ended strategies for learning. For example, computer graphics is revolutionizing the study of shape. Just as the supercomputer is changing methods of research, so ordinary computers are providing images that most of us could not imagine a decade ago.

Many teachers say that computer software has completely changed the way they teach. They no longer feel that they have to have all the answers; instead, they become partners with the students in exploring the properties of shape. These teachers are very enthusiastic about their new way of teaching. Both their enthusiasm and the new "partnership pedagogy" can be encouraged by imaginative curricula that embed exploration of shape throughout the entire curriculum.

REFERENCES AND RECOMMENDED READING

1. Arnheim, R. *Visual Thinking.* Berkeley, CA: University of California Press, 1969.
2. von Baeyer, H.C. "A dream come true." *The Sciences* (New York Academy of Sciences), (Jan.–Feb. 1989), 6–8.
3. Berger, Marcel. *Geometry.* New York, NY: Springer-Verlag, 1987.
4. Coxeter, H.S.M. *Introduction to Geometry.* New York, NY: John Wiley & Sons, 1969.
5. Coxeter, H.S.M. *Regular Polytopes.* New York, NY: Dover, 1973.
6. Edgerton, Samuel Y., Jr. "Galileo, Florentine 'Disegno,' and the 'Strange Spottednesse' of the Moon." *Art Journal,* 44 (1984), 225–232.
7. Friedlander, Alex and Lappan, Glenda. "Similarity: Investigations at the Middle Grade Level." *Learning and Teaching Geometry, K–12.* Reston, VA: National Council of Teachers of Mathematics, 1987, 136–143.
8. Galileo. *Sidereus Nuncius (The Starry Messenger),* 1610.
9. Gombrich, E.H. *The Sense of Order.* Ithaca, NY: Cornell University Press, 1979.
10. Grünbaum, Branko and Shephard, G.S. *Tilings and Patterns.* New York, NY: W.H. Freeman, 1987.
11. Grünbaum, Branko. "Geometry strikes again." *Mathematics Magazine,* 58:1 (1985), 12–18.
12. Holden, Alan. *Orderly Tangles.* New York, NY: Columbia University Press, 1983.
13. Markoff, John. "Supercomputer pictures solve the once insoluble." *The New York Times* (Oct. 30, 1988), 1, 26.
14. National Council of Teachers of Mathematics. *Learning and Teaching Geometry, K–12.* Reston, VA: National Council of Teachers of Mathematics, 1987.
15. National Council of Teachers of Mathematics. *Curriculum and Evaluation Standards for School Mathematics.* Reston, VA: National Council of Teachers of Mathematics, 1989.
16. Senechal, Marjorie and Fleck, George. *Patterns of Symmetry.* Amherst, MA: University of Massachusetts Press, 1977.
17. Senechal, Marjorie and Fleck, George. *Shaping Space: A Polyhedral Approach.* Boston, MA: Birkhauser, 1988.
18. Senechal, Marjorie and Fleck, George. *The Workbook of Common Geometry.* (In preparation)
19. Senechal, Marjorie. "Symmetry revisited." In Hargittai, Istvan (Ed.): *Symmetry II.* Elmsford, NY: Pergamon Press, 1989.

20. Stevens, P. *Patterns in Nature.* Boston, MA: Little, Brown & Company, 1974.
21. Thompson, D'Arcy W. *On Growth and Form,* Abridged Edition. Cambridge, MA: Cambridge University Press, 1966.
22. Watson, James and Crick, Francis. "Structure of small viruses." *Nature,* 177 (1956), 473–75.
23. Weeks, Jeffrey R. *The Shape of Space.* New York, NY: Marcel Dekker, 1985.

Change

~ ~

IAN STEWART

Every natural phenomenon, from the quantum vibrations of subatomic particles to the universe itself, is a manifestation of change. Developing organisms change as they grow. Populations of living creatures, from viruses to whales, vary from day to day or from year to year. Our history, politics, economics, and climate are subject to constant, and often baffling, changes.

Some changes are simple: the cycle of the seasons, the ebb and flow of the tides. Others seem more complicated: economic recessions, outbreaks of disease, the weather. All kinds of changes influence our lives.

It is of the greatest importance that we should understand and control the changing world in which we live. To do this effectively we must become sensitive to the *patterns* of change, including the discovery of hidden patterns in events that at first sight appear patternless. To do this we need to:

- *Represent* changes in a comprehensible form,
- *Understand* the fundamental types of change,
- *Recognize* particular types of changes when they occur,
- *Apply* these techniques to the outside world, and
- *Control* a changing universe to our best advantage.

The most effective medium for performing these tasks is mathematics. With mathematics we build model universes and take them apart to see how they tick, we highlight their important structural features, and we perceive and develop general principles. Mathematics is the ultimate in "technology transfer": patterns perceived in a single example can be applied across the entire spectrum of science and business.

THE MATHEMATICS OF CHANGE

The traditional approach to the mathematics of change can be summed up in one word: *calculus.* In calculus the changing system is modeled by a special equation (technically, a *differential* equation) that describes the relation between the rates of change of different variables. As much heavy machinery (both theoretical and numerical) as is required is brought to bear in an effort to solve the equation. Preparing students for the study of calculus has been the central goal of school mathematics; setting up and solving the equations of calculus is the lifeblood of traditional engineering mathematics.

Calculus remains an essential component of the mathematics of change. Newer methods such as discrete mathematics and computation enhance rather than replace calculus. But mathematics is itself subject to change. New problems and new discoveries imply the need for a much more varied range of mental equipment. Two important trends are worth mentioning: the use of increasingly sophisticated approximate methods and exploitation of geometry and computer graphics. The first has been made possible by the enormous increase in computer power. Because computing is based on digital manipulation, it requires an understanding of the discrete as well as the continuous—and above all, of the relation between the two.

The second trend is a major triumph of mathematical imagination: the use of visual imagery to condense a large quantity of information into a single comprehensible picture. Computer graphics has led to the discovery that many aspects of change are manifestations of a relatively small number of fundamental geometric forms. Mathematicians are just beginning to understand these basic building blocks of change and to analyze how they combine. The methodology involved has a very different spirit from traditional modeling with differential equations: it is more like chemistry than calculus, requiring careful counterpoint between analysis and synthesis.

The graphical representation of various mathematical concepts arising in the study of change has led to the discovery of a variety of intricate shapes, each of which appears in many different dynamical situations and is thus a "universal" object in the mathematics of change.[14] Figure 1 portrays a number of these shapes. They illustrate well the vast differences between today's visual methods and the forms traditionally

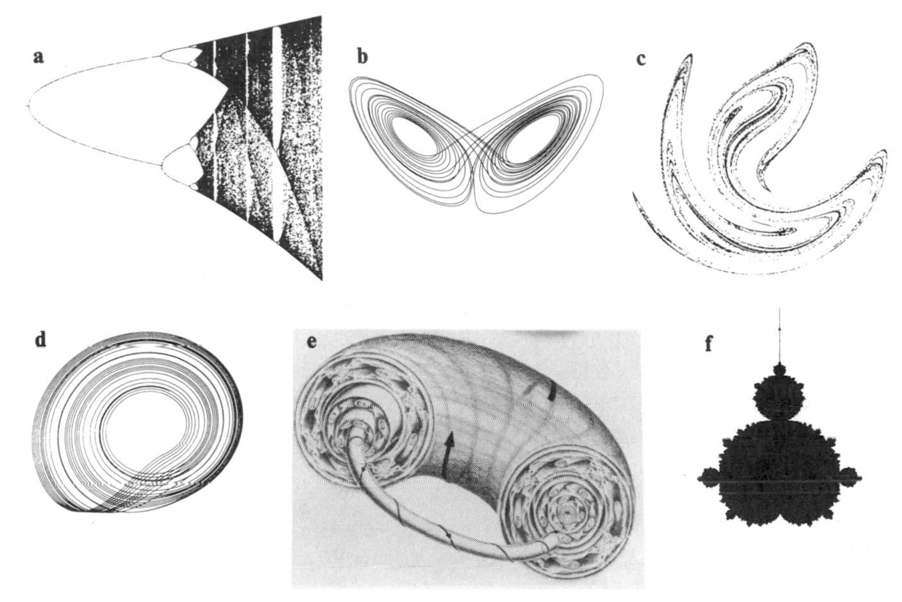

FIGURE 1. New scenery in the landscape of change: (a) period-doubling cascade, (b) Lorenz attractor, (c) Ueda attractor, (d) Rössler attractor, (e) vague attractor of Kolmogorov, (f) Mandelbrot set.

studied in geometry, such as triangles and parallelograms.[1,17] Geometry is now organic and visual rather than limited and formal.

In consequence, there are very few branches of mathematics today that do not have some bearing on change. In part this is because mathematics is a highly integrated and interconnected structure. Furthermore, change is such a complex and varied phenomenon that we need all the ideas we can muster to handle it. To study change the scientist of the future will need to combine, in a single integrated world view, aspects of traditional mathematics, modern mathematics, experimentation, and computation. We will need scientists who reach as readily for a pencil as for a computer terminal, who can draw crude but informative sketches as readily as a computer graphic, and who think in pictures as readily as in numbers or formulas. The entire point of view—the mental tool kit—of the working scientist will be very different from what it was even a decade ago.

The patterns of change in nature and in mathematics are unconstrained by conventional categories of thought. In order to make progress we must respond imaginatively and sensitively to new types of pattern. Our own patterns of thought must themselves change.

Variety of Styles

As the twentieth century draws to a close, a new style of mathematics is emerging—a style whose characteristic is variety. Mathematics is once again developing in close conjunction with its applications to science—physical, biological, behavioral, and social. Much mathematics is inspired by computer or laboratory experiments or by the forms of natural phenomena. Conversely, mathematical ideas developed for their own sake, or in some distinct area of application, are being transferred to other tasks and put to work.[10,25] This variety is a strength of the new style of mathematics, and it should be encouraged at all levels. Moreover, computers (especially computer graphics) allow nonspecialists—from school children to managers, from school teachers to scientists—to witness the beauty and complexity of mathematics and to put it to work.[3,17]

The emergence of this new style of mathematics does *not* imply that the traditional emphasis on precise formulation of concepts and rigorous logical proof can be abandoned. On the contrary, they remain an essential component of the mathematical endeavor. Rigor and precision are as essential to mathematics as experiment is to the rest of science, and for much the same reason: they provide firm reasons for believing that ideas and methods are sound. They are part of the subject's internal checks and balances, a constant safeguard against error. The training of professional mathematicians will necessarily continue to require accurate logical thinking and a precise understanding of the meaning of "proof." The use of computers as "experimental tools" in mathematics can stimulate and motivate new ideas and problems, but these experiments alone cannot provide understanding of *why* the observed phenomena happen. Their role is to offer a degree of confidence that certain phenomena do indeed occur.

In fact an important trend has become very noticeable, as experience in the use of computers has developed. It is the disappearance of the dismissive attitude, "Put it on the computer and that will answer all your questions." When the answer to a problem is, say, a single number, such as the failure load of an engineering structure, all of one's problems indeed do disappear once that number is known. But today a typical computer-based investigation may produce several hundred diagrams representing the behavior of the system under various conditions. For example, think of the flow of air past a space shuttle for different speeds, angles of attack, and atmospheric densities. Such a catalogue, despite its apparently large size, is likely to be inadequate for determining the behavior under *all* possible conditions. If the system involves three adjustable parameters, as does the one just mentioned, and each can take up to ten values, then a total of a thousand combinations is possible. With four such variables there are ten thousand, with six there are a million.

In practice, six is a small number of parameters: simple problems in chemical engineering typically involve several dozen parameters and may involve hundreds. It is pointless to produce a computerized catalogue of one million diagrams, let alone a billion or a trillion. The fundamental question—"What is *really* going on here?"—returns from computer science to the realm of mathematics. Such questions require input from the human brain far more than from the computer.

However, the role of the computer should not be underestimated. It is becoming an ever more prevalent thinking aid. Computers cannot only generate "results," but they can also be used to experiment at intermediate stages of understanding, to test hypotheses and possible mechanisms. With appropriate safeguards, computer calculations can actually produce rigorous proofs of mathematical results. Such computer-aided proofs require very careful construction and a great deal of human input to set them up: they are far from routine and usually require specially constructed software and lengthy machine time. More than anything else, they constitute a difficult specialist area of mathematics. "Put it on the computer" is no panacea.

Approaches to Teaching

For reasons of exposition only, rigorous proof does not feature prominently in this essay. It is part of the mathematician's basic technique, and it remains just as important as it ever was, but it holds much less interest for the nonspecialist. Accordingly, its role has not been made explicit, although it underpins everything discussed.

However, the fact that proof is important for the professional mathematician does not imply that the teaching of mathematics to a given audience must be limited to ideas whose proofs are accessible to that audience. Such a limitation is likely to make mathematics dull, dry, and dreary, for many of the most stimulating and exciting ideas depend upon highly complex theories for their proofs. Many mathematical concepts can be grasped without being exposed to their formal proofs. Using an idea is quite different from developing it. It is possible to "explain" quite advanced concepts to children by means of examples and experiments, even when a formal proof is too difficult.

For example, in the theory of chaos an important concept is that of "sensitivity to initial conditions." If a system evolves from two very similar initial states, the resulting motions can quickly become totally different. Given access to suitable software, virtually anyone can appreciate this sensitive and paradoxical behavior in, say, the Lorenz attractor (Figure 1b) merely by watching how two almost equal starting values move apart and become independent. However, a rigorous proof that the Lorenz system really does behave in the manner that computer experiments suggest is not only beyond the capacities of the average person, it has not yet been achieved by professional mathematicians and remains an active problem for future research.

The breadth of viewpoint and range of skills demanded by today's mathematics will be important, not just for mathematicians and scientists but for people in all walks of life. Change affects us all. Managers, politicians, business leaders, and other decision makers must cope with a changing world. They must appreciate how subtle change is; they must unlearn outdated assumptions.

It is a tremendous challenge to devise methods of educating a generation of such versatile people. Our aim here is to suggest ways to develop in children some of the underlying ideas and to stimulate a new point of view. We must advance beyond the traditional approach of arithmetic leading to algebra and thence to calculus.

In the design of an effective new curriculum, one important component is an understanding of the new viewpoints that are developing at the frontiers of research. Yet the curriculum must be suitable for all children, not just for those who will become research scientists. Nevertheless, new kinds of mathematics that are evolving at the research level set the style for applications and education in the future. Thus it is important for teachers and educators at all levels to understand the general nature of these new methods and the kinds of questions that they address.

Levels of Description

The mathematics of change can be viewed at many levels:

- The big picture: What are the possible types of change?
- Specific areas of mathematical technique: How are the equations solved?
- General areas of application: How does the size of an animal population vary with time?
- Individual applications: Design a chemical reactor to produce margarine.
- Simple theoretical examples: How does a pendulum oscillate?

Mathematicians operate on all of these levels because insights obtained at one level are often transferred to other levels. In mathematical technology transfer, patterns are not tied to any particular area of application.

Simple theoretical examples are seldom of direct relevance to industrial applications. For example, an analysis of pendulum dynamics is of no direct use in the study of wing flutter in supersonic aircraft. In practical terms the pendulum went out with the grandfather clock. But simple examples have their uses: they prepare us for the complexities of real life. A pendulum makes many important features of oscillation more accessible than would a realistic model of a vibrating airplane wing.

To illustrate these themes we will use some specific questions that exemplify the new style of mathematics. These questions have been chosen not as specific goals in themselves, but because they motivate compelling mathematical ideas:

- How do living populations change?
- Where do meteorites come from?
- Why are tigers striped?

Only the first of these questions appears to involve change. The others seem to be about static phenomena.[27] Meteorites are just there—or not—at random. A tiger is striped, a leopard is not, and never the twain shall meet. In fact the questions are all about change of some kind. Do meteorites really plunge into the earth's atmosphere "at random," or does something more structured lie behind their appearance in the night sky? A mature striped tiger does not just exist as a static object: it develops from a single (unstriped) cell. Somewhere along the line of development the stripes first make their entrance. Change is the common theme behind each of these varied questions.

POPULATION DYNAMICS

If we put a few rabbits on an uninhabited island, pretty soon there will be a lot more rabbits. On the other hand, the growth cannot continue unchecked, or soon there would be more rabbits than island. It follows that change in a population is affected by both internal and external factors. How they combine to influence changes in the population is a good example of mathematical modeling that can be studied at many different levels.

Limits to Growth

We begin with the simplest case: a population consisting of a single species with a constant (and therefore limited) food supply. Figure 2 shows typical experimental data for growth of such a population. Its typical S-shaped curve is characteristic of many growth phenomena.[27]

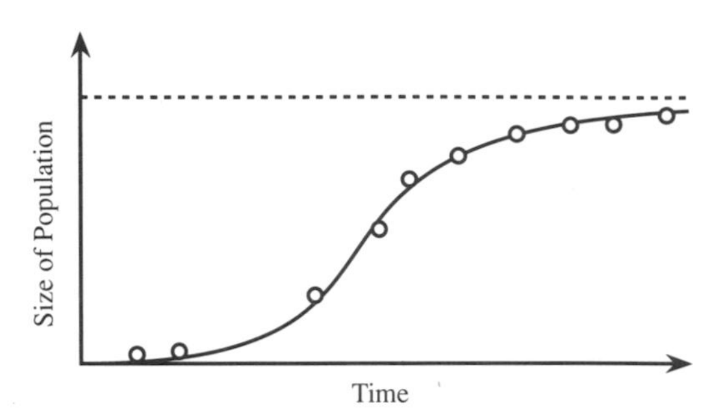

FIGURE 2. Changes in the size of a yeast population growing in an environment with a limited food supply.

Similar curves arise if we measure particular features of a single developing organism—for example, the height or weight of a growing child.

It is common in many families to record the heights or weights of children as they grow. These charts may be displayed on classroom walls for discussion and comparison. The growth of young children in a single class over a period of one or two years will illustrate *linear* growth. The heights on the chart, plotted against time, will lie close to a straight line. However, the complete growth record of a child from birth to adulthood exhibits the characteristic *S* shape. Neither the initial phase nor the final phase is linear. Early on the growth is approximately exponential; later it saturates as it approaches a constant value.

Children who have recorded growth curves can be introduced to the entire *S*-shaped curve, either as an experimental observation or as a table of numbers. A good exercise for children in middle school is to use evidence from several childrens' growth curves together with data from their own childhood to project their own adult heights. Later, as older students, they can learn how to represent these curves with formulas. Children can be encouraged to analyze the main features of this curve and to consider *why* the curve has them.

Suppose Alice is 1 foot tall at age 0 and 4 feet tall at age 8. If this growth rate continues—3 feet every 8 years—how tall will she be at ages 16 or 24 or 32? (Answers: 7 feet, 10 feet, 13 feet.) Even young children can see that these answers are not credible. What's wrong? The mathematics is fine, but the model—linear growth—is inappropriate. Moral: When you use mathematics you have to pick a sensible model and not just calculate numbers blindly.

Levels of Analysis

A study of population growth can be carried on at several levels—verbal, numerical, graphical, dynamical—with the sophistication increasing as the children become older. Verbal description of the yeast growth curve shows a population that increases slowly at first but then grows exponentially. That is, the breeding population increases by a constant factor in successive periods of time. However, when the population becomes sufficiently large, the rate of growth slows down, eventually leveling off at a steady maximum value.

This verbal model is purely descriptive. It is mute about why the population levels off. The verbal description is helpful for general intuition but useless for further analysis of behavior. Its principal role is to summarize simply the pattern of growth.

To appreciate the effect of exponential growth—and to gain insight into why such growth cannot continue unchecked—children can be told the famous story about the emperor's reward. In a far country a person performed an important task for the emperor, and she was asked to name her reward. The reply was: "One grain of wheat on the first square of a chessboard, two on the next, then four on the next, then eight, and so on, doubling each time." The emperor was not very impressed ...until he worked out how the numbers grew!

Children can do the same with a calculator or a computer. Younger children can experience exponential phenomena without using large numbers by folding a sheet of paper repeatedly in half. How many times can you manage before you get stuck?

Data on weights of animals, wingspans of birds, girths of trees, numbers of leaves on plants, etc., can be gathered (or presented) in the form of numerical tables. Children can look for patterns in the numbers: Are they increasing? Decreasing? Constant? They can calculate differences and ratios, make tables, and look for patterns. Numerical tables lead naturally to graphical representation.

The growth curve provides a visual picture of the way in which the two variables, population and time, are related. Such a graph, sometimes called a time series, replaces numerical information by graphical: it is the simplest example of the geometrization of change. The idea that numbers can be represented by the positions of points, and changing numbers by curves, is the basis of all geometric methods in the mathematics of change (see Figure 3). Children need many opportunities to learn that in mathematics a picture is indeed worth a thousand words.

For younger children experimental work is most appropriate. They can count the number of eggs produced by ducks or chickens, measure the height of a growing plant, measure the temperature each day at noon, record the position of the moon in the sky. By graphing this data, children can search for patterns of change and discuss possible causes.

Older children can be set more ambitious tasks: the water level in a pond, the number of leaves on a bush, the movements of the stock market, experiments from physics and chemistry laboratories. Using data from real phenomena is an effective way to integrate mathematics into other school subjects. Algebra students can also use mathematical

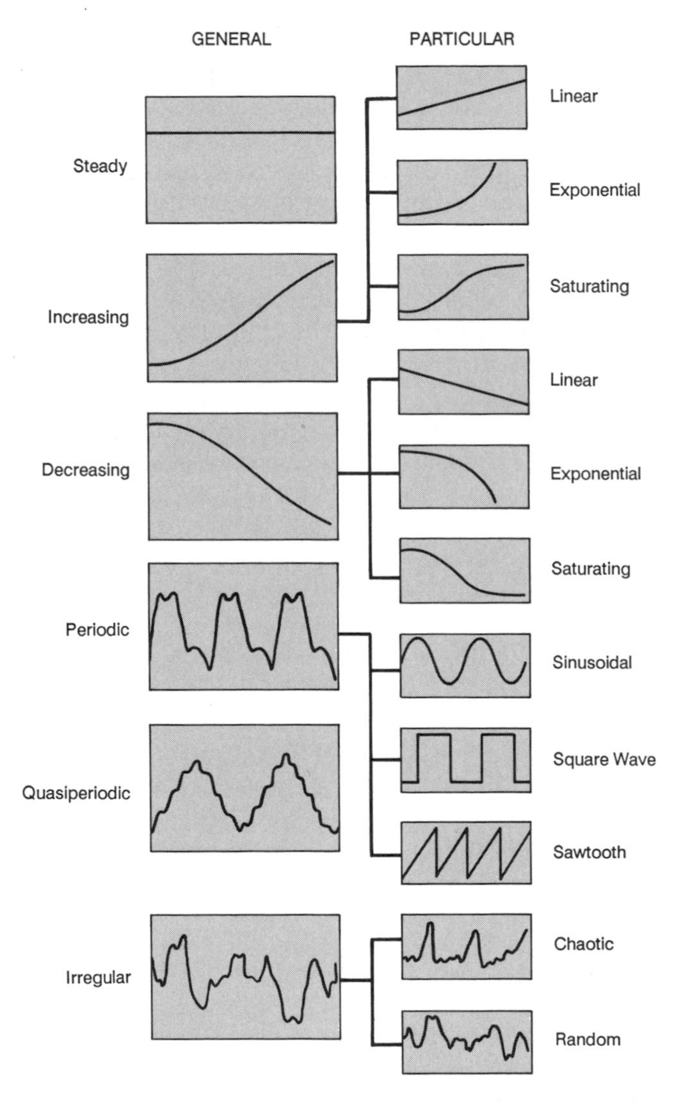

FIGURE 3. Some of the many different types of change, together with their typical time series.

processes and formulas to generate theoretical data, to look for patterns, and to compare theory with reality.

Dynamical Systems

The next level of exploration is to model not the patterns in the numbers but the *process* that gives rise to these patterns. In the traditional approach this idea leads to differential equations and thus requires calculus. But another possibility—increasingly attractive in an age of computers—is to throw off the chains of calculus and take seriously the fact that the number of creatures in a population is discrete rather than continuous.

Imagine that time t increases in *discrete* whole number steps, $t = 1, 2, 3, \ldots$. The value of the population p at time t is written as $p(t)$. Its next value $p(t+1)$ can then be related to its current value $p(t)$ by a specific growth law. This type of model is called a *difference* equation or a discrete system.[7,28]

In living populations, unchecked breeding at a constant rate m corresponds to a law of the form $p(t+1) = mp(t)$, leading to exponential growth: $p(t) = p(0)m^t$, where $p(0)$ is the initial population. The law of restrained growth, which allows for limits imposed by lack of food or space, modifies this law by subtracting a correction factor that reflects these limits:

$$p(t+1) = mp(t) - n[p(t)]^2 ,$$

where m and n are constants that depend on the particular circumstances. This equation, known as the Verhulst law (named after the nineteenth-century French scientist P.F. Verhulst), is one of the most common algebraic models of limited growth.

Students can study this equation with tools from simple algebra, both by making tables and by simplifying the equation. The population level $p(t) = m/n$ is a cutoff level: once it is reached, the next value, $p(t+1)$, is 0, as are all subsequent values. To study how the population compares with the cutoff level, we can express $p(t)$ as a proportion of m/n by changing the units of measurement by letting $p(t) = q(t) \cdot (m/n)$. This leads to the equation

$$q(t+1) = m(q(t) - q(t)^2),$$

where $q(t)$ expresses the population as a fraction of the cutoff level. Instead of two parameters m and n, we now have just one parameter m, which makes the mathematics much simpler. Because $q(t)$ is a proportion of the cutoff population, it will be some fraction between 0 and 1.

Difference equations such as the Verhulst law are ideal for computer calculation, because they express a simple repetitive procedure for describing the behavior at the next instant $t+1$ from the behavior now, at time t. With a computer we can easily calculate solutions of the discrete Verhulst law without knowing a formula for these solutions.

(Indeed, there *is* no general formula for these solutions.) We can then encapsulate the results in a single geometric object such as a time series graph.

This illustrates an important general principle: discrete mathematics is often more accessible than continuous mathematics (calculus). The Verhulst law can be introduced and studied through tables of values as soon as students begin their study of algebra, usually four years before they are introduced to calculus. However, it is also harder to derive the detailed mathematical structure of discrete systems, and their treatment tends to be experimental or at least computer based.

Numerical Experiments

The Verhulst law offers an excellent opportunity for numerical experiments using only elementary arithmetic and calculators.[3,23] Even elementary school children can follow the rules, years before they are introduced to the formalism of algebra. The Verhulst law, whose algebraic form is

$$p(t+1) = m[p(t) - p(t)^2]$$

can easily be translated into a table or a spreadsheet for exploration for various values of the parameter m. (Note that we are now using p to signify the population proportion, which we previously called q, rather than the population size.)

Start with some value of $p(0)$, say 0.1, and calculate in turn $p(1)$, $p(2)$, $p(3)$, ... In words: new population equals old population minus the square of the old population, multiplied by a constant.

For example, suppose $m = 2$. Then the successive values are

0.1, 0.18, 0.295, 0.416, 0.486, 0.499, 0.5, 0.5, ...

We see initial growth, settling down to a specific final level. This growth is similar to the experimentally verified growth of yeast and other homogeneous populations (see Figure 2). When $m = 3$ we get

0.1, 0.27, 0.591, 0.725, 0.598, 0.721, 0.603, 0.717, ...

The values in this case appears to oscillate between about 0.6 and 0.7. (In fact, this oscillation eventually dies out, but very slowly: it becomes more apparent at $m = 3.1$ or 3.2.) Finally, consider $m = 4$:

0.1, 0.36, 0.922, 0.289, 0.821, 0.585, 0.970, 0.113, ...

Now we see no clear pattern at all! What has happened?

The Verhulst law leads to a rich range of behavior, including periodic oscillations and apparently patternless, irregular behavior. The latter is known as *chaos*. Here a simple experiment using a calculator brings quite young children to the frontiers of research. Indeed, this example can lead to an enormous range of classroom activities: working out numerical values on calculators, computers, or electronic spreadsheets; graphing the results; spotting patterns; analyzing why they occur.

Traditionally, random-looking behavior is modeled by statistics, using equations that incorporate explicit random terms. But there is no random term in the Verhulst law: it is *deterministic.* This example shows, surprisingly, that behavior predicted by a simple and explicit law can be highly irregular, even random.

This paradoxical discovery is called *deterministic chaos.* Irregular fluctuations may arise from nonrandom laws, making it possible to model many irregular phenomena in a simple manner. It also demonstrates that simple causes can produce complicated effects. It is one of the most exciting areas of current mathematical research.[5,11,24]

The Irregular Fruit Fly

It is always possible that chaotic behavior could be just an artifact of the model and not a phenomenon of nature. Perhaps. But natural populations do, in fact, display irregular oscillatory behavior. Figure 4 shows experimental data on a population of fruit flies kept in a closed container and fed a constant protein diet.[15] When the population rises too high, there is too little food and the flies are unable to breed properly. The population then drops until there is excess food; then the flies breed unrestrictedly and the population shoots up again.

The main overall effect is an oscillation with a period of about 38 days. However, as the time series shows, the way in which the population changes is decidedly complex. Many of the peaks in the graph are double, being more M-shaped than Λ-shaped. The height of the peak

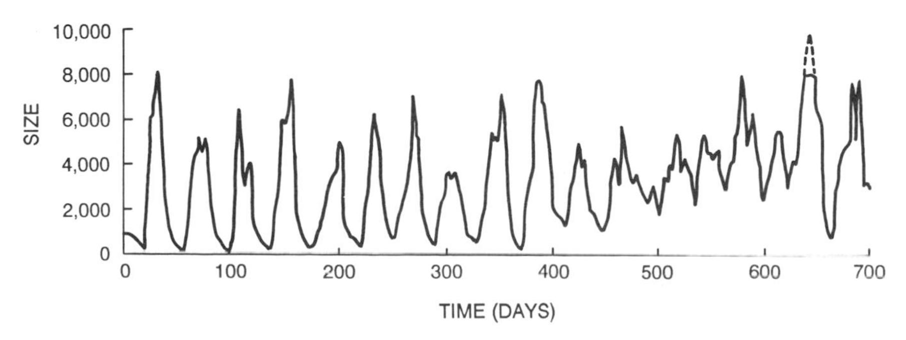

FIGURE 4. Variations in an experimental population of fruit flies show irregular oscillatory behavior that is typical of deterministic chaos.

varies: small, medium, large, in turn. After the first 450 days or so, the changes become more and more irregular.

This graph illustrates an important question for mathematical modeling and for the analysis of scientific data. Some of the observed changes are due to the population dynamics of fruit flies. Others may be due to outside effects such as contaminated food, disease, or—for all anyone knows—the tides or the position of Mars in the sky. How can we tell which are which?

It would be easy to assume that the regular effects—the M-shaped peaks, the modulation in their size—are unrelated to outside causes but that the increasing irregularity after 450 days is due to something going wrong with an outside cause. However, this assumption may be incorrect. Numerical experiments with models similar to the Verhulst law show that simple mathematical laws can produce both regular oscillations and irregular chaos, just by making slight changes to a single parameter. In fact many aspects of the fruit fly data, irregularities included, can be modeled by simple systems.

Children can be brought to understand the possibilities for complex behavior in simple systems by performing numerical experiments, first with calculators and later with computers. They can then search for patterns in apparently irregular data. For example, given a time series generated by the Verhulst law or related equations, they can plot $p(t+1)$ against $p(t)$ and observe that all the points lie on a smooth curve. They can analyze the curve to determine its geometric features; older children can seek an appropriate formula and estimate the value of the growth rate parameter m.

More sophisticated versions of this geometric technique have been applied to many sets of observational data, for example, to the apparently random fluctuations that occur in the numbers of people suffering from a disease such as measles. Often the experimental time series appear random. But graphical analysis suggests that a simple process, resembling a difference equation, underlies the apparent irregularities. In consequence it is often possible to set up simple but realistic models that reproduce the patterns of change in these systems.

Moving on to Calculus

Traditional analysis via calculus still has an important role to play in modeling population growth. In this case it provides a *formula* rather than a picture or a list of numbers. In the calculus-based model the value of $p(t)$ need not be a whole number, whereas a real population necessarily takes on whole number values. The model is thus a continuous approximation to a discrete phenomenon. This is a common technique and is often used when the maximum population size is fairly large. Then the change caused by adding or removing a single individual is extremely tiny, so that the possible range of sizes cannot easily be distinguished from a continuous range. The resulting model is a *differential equation,* one of the key concepts of higher mathematics.[4] A differential equation involves not just variables such as the population p but also rates of change of variables. The rate of change of a variable p with respect to time is traditionally denoted by dp/dt.

The simplest differential equation for populations is a law of uniform growth $dp/dt = mp$. This states that the rate of change dp/dt of the population p at a given moment t is proportional to the population p at that same moment, where the constant of proportionality is m. In other words, a larger population produces proportionately more offspring than a smaller one. The solution to this differential equation is $p(t) = p(0)e^{mt}$ for an initial value $p(0)$ at $t = 0$, which is the continuous version of exponential growth. The population explodes, unchecked.

In practice other factors must come into play to limit the growth. As with the Verhulst law, we modify the equation by subtracting a term np^2 (where n is a second constant):

$$dp/dt = mp - np^2.$$

The point of this extra term is that when p is small, p^2 is negligible in comparison, so that the correction term np^2 has little effect; in this case we obtain (almost) exponential growth. However, as p becomes larger, the term $-np^2$ begins to dominate the dynamics, substantially reducing the rate of growth. Indeed when p reaches the value m/n, the rate of change of the population, dp/dt, becomes zero. When this happens, no further growth takes place. So m/n represents the maximum population. Using techniques of calculus, it is possible to find a formula for the solution. The graph of this solution, known as the *logistic curve,* has the same S-shape as the experimental data on yeast (Figure 2).

The rich variety of behavior—steady, periodic, chaotic—of the discrete Verhulst law is absent from its continuous analog, which yields only a smooth S-shaped curve. This shows in a particularly convincing manner that changing from discrete models to continuous ones, or conversely, can lead to new phenomena: it is not just a harmless trick. Examples such as these raise important questions about the relation between continuous and discrete models, relations worth exploring in mathematics classes at many school levels.

The continuous model permits experimental data to be fitted to a theoretical curve, and this opens the way to prediction of future behavior. For example, if a logistic curve is fitted to the population of the United States up to 1930, it predicts that by the year 2000 the population should level off at around 200 million. More accurate techniques give a projected population for the year 2000 of 260 million, about 30% higher. So the simplified approach does surprisingly well. Students armed with population data (of an ecosystem, a nation, or the world) can try fitting the logistic curve to this data to determine the constants m and n and to predict future trends.

METEORITES

The behavior of meteorites is a small part of the general problem of the dynamics of celestial bodies—of moons, planets, stars, galaxies. The regularities, or almost regularities, of the motions of the planets have throughout history been a major motivation for the study of change. It is not just a matter of fascination with the night sky: important down-to-earth problems such as agriculture and navigation have at various times depended upon knowledge of the movements of the stars and planets.

Astronomy is a rich area for finding good classroom activities about change: the phases of the moon, the tides, the apparent motion of stars, the changing seasons, earth satellites. Another possibility is to reconstruct Galileo's experiments using balls on inclined slopes and deduce the law of motion in a uniform gravitational field. Data gathered in such enterprises can fuel many rich mathematical explorations.

Historically, our understanding of such matters went through several stages—informal description, empirical models, geometrical models, dynamical models—before culminating in the laws of motion discovered by Isaac Newton. But these laws often lead to equations that are very hard to solve. They can be solved exactly for a system of two bodies, where they predict elliptical orbits. The problem of celestial motion for a system of three bodies has been notorious for over two centuries for its apparent intractability. With modern computers we can see why: even simplified versions—for example, where one body has negligible mass—lead to complex and highly irregular behavior.

Computer packages now simulate planetary motion for systems of two, three, or more bodies. Children as young as 11 or 12 can use these packages to experiment with the behavior of the regular elliptical orbits of two-body systems and the complicated behavior of three or more bodies. By using these packages, they can gain more insight into the geometry of planetary motion than Isaac Newton did in a lifetime of study.

Stability

Modern understanding of planetary motion stems from work of the French mathematician Henri Poincaré around the turn of the century.[9,22] In 1887 King Oscar II of Sweden offered a prize of 2500 crowns for an answer to a fundamental question in astronomy: Is the solar system stable? We see now that Poincaré's response was a major turning point in the mathematical theory of celestial change.

Scientists call a system *stable* if it does not change when perturbed by small disturbances. It is unstable if small disturbances tend to become magnified, leading to large changes in behavior. For example, a pin lying on its side is stable, whereas a pin balanced on its tip is unstable since it will always fall over (Figure 5).

Children can develop sound intuition about the notions of stable and unstable systems, and indeed for the typical complexity of dynamical systems, by exploring the behavior of various mechanical "executive toys"—multiple pendulums, interacting magnets, gyroscopes. For example, consider a pendulum with a magnetic bob, arranged to swing over the top of a second magnet. If the two magnets have opposite polarity, then the pendulum is stable in its downward position, attracted by the lower magnet. But if the polarities are the same, and you try to hold the pendulum over the lower magnet, it tries to move away. The downward position is now unstable—and the child can *feel* it!

Experiments of this type, usually carried out in a rather formal way, are currently characteristic of physics classes. Less formal experiments should be carried out in mathematics classes as well, as an integrated part of the development of intuition for change and motion, for stability and chaos. At later stages, after a child's intuition is better developed, such experiences can be formalized with appropriate mathematical models.

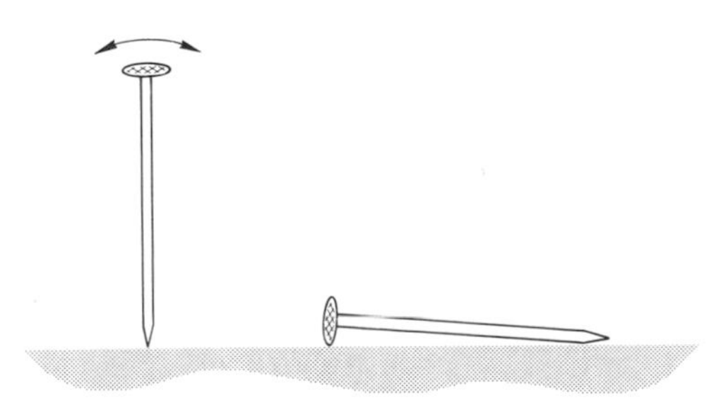

FIGURE 5. Unstable and stable states of a pin: when balanced on its tip, any wiggle will cause a pin to fall, whereas when resting on its side, small forces produce only small changes in the position of the pin.

Stability is an extremely important question. An airplane must not only fly, but its flight must be stable, or it will drop out of the sky. When a car rounds a corner it must not tip over on its side. The solar system is a very complicated piece of dynamics. How do we know that the motion is stable? Will all the planets continue to move in roughly their current orbits? Could Pluto crash into the sun? Could the earth wander off into the cold of the outer planets? These are very subtle problems whose answers are very difficult to discern.

Rubber Sheet Dynamics

Poincaré didn't solve King Oscar's problem: it was too hard. But he made such a dent in it that he was awarded the prize anyway. To do it he invented a new branch of mathematics now called *topology*. Often characterized as "rubber sheet geometry," topology is more properly defined as the mathematics of continuity, as the study of smooth, gradual changes, the science of the unbroken.[8,18] Discontinuities, in contrast, are sudden and dramatic—places where a tiny change in cause produces an enormous change in effect.

The celestial motion of two bodies—a universe consisting only of the earth and the sun, say—is periodic: it repeats over and over again, once every year. (That is the definition of "year.") This periodic behavior immediately proves that in such a solar system—containing only the earth and the sun—the earth would not fall into the sun or wander off into the outer reaches of infinity; for if it did, it would have to fall into the sun every year or wander off to infinity every year. Those aren't things you can do more than once, and they didn't happen last year, so they never will. In other words, periodicity gives a very useful handle on stability. In our real universe bodies will disturb this simple scenario; nevertheless, periodicity is still important.

Under gravity, two bodies behave simply: they both move in elliptical orbits about their common center of gravity. Three bodies behave in an unbelievably complicated manner, even if the problem is simplified by assuming that one has a very small mass compared with the other two. More than three bodies can lead to even worse behavior.

Juggling is an example of stable periodic motion. It is periodic because the same actions are performed over and over again; and it must be stable since otherwise it wouldn't work. Juggling two bodies is relatively simple; juggling more quickly becomes very complicated. If one teaches children to juggle, they will learn quickly about the complexity of dynamical systems. They can analyze the periodic pattern of juggling motion. Why is juggling stable? What is the role of hand-eye feedback?

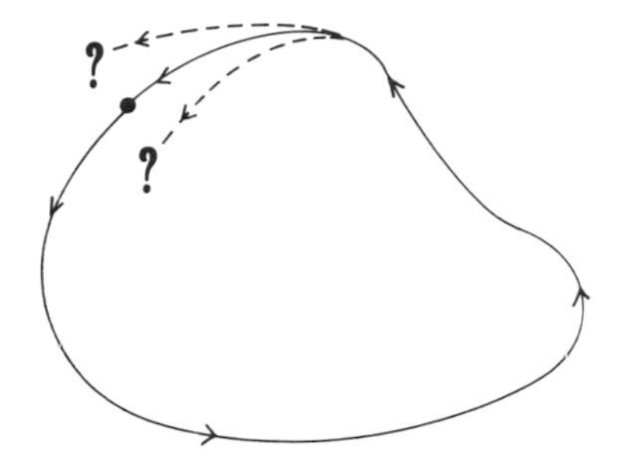

FIGURE 6. Poincaré's geometric approach to periodicity: if the state of a system describes a closed loop in phase space, the system must be periodic and hence stable.

Poincaré grappled with the existence of periodic solutions, and he found that they could be detected by a topological method. Suppose that at some particular instant of time the system is in some particular state and that at a certain time later it is again in the *identical* state. Then it must repeat, over and over again forever, the very motion that took it from that state back to itself. Returning just once to a previous state, perfect in every detail, is the essence of periodic motion.

Topology enters when this idea is made geometric.[24] Imagine that the state of the system is described by the coordinates of a point in some high-dimensional space, which scientists call *phase space.* As the system changes, this point will move, tracing out a curve in phase space. In order for the system to return to its initial state, this curve must close up into a loop (Figure 6). Stability of the system thus translates to "When does a curve form a closed loop?" The question asks nothing about the shape or size or position of the loop, merely that it be closed: it's a question for topology. Thus the existence of periodic solutions depends on topological properties of the curve that represents the changing state of the system in phase space.

Phase space is an abstract mathematical space with many dimensions that represent all possible variables that govern the state of a system,

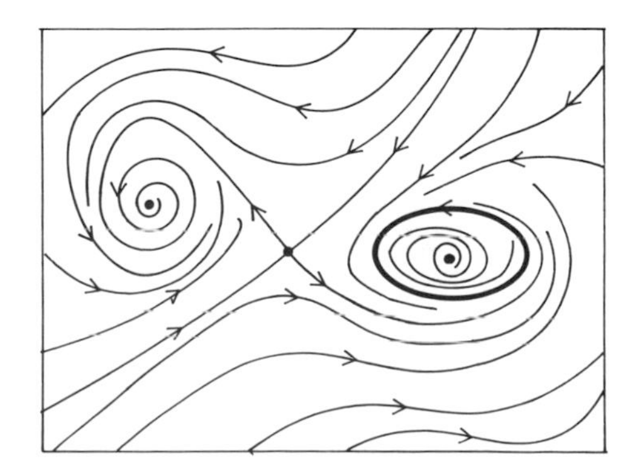

FIGURE 7. Example of a phase portrait in which different curves represent possible evolution of a system under different initial conditions.

which is itself represented as a single point in phase space. As the state changes, this point moves, tracing out a curve, or flow line. The picture of how these flow lines fit together is called the *phase portrait* of the system.[1] The flow is typically indicated by curved lines, corresponding to the time evolution of the coordinates of various initial points (see Figure 7). Arrows mark the direction of motion of time.

Phase Portraits

Once children have grasped the concept of graphing the changes in a single variable, they can be introduced to phase portraits. Instead of plotting the value of a single variable against time, in a time series they can plot the sequence of values of two different variables in two coordinate directions. Such exercises will develop insight into the multidimensional geometry of change. For young children these variables might be the height and weight of growing animals or the temperature and rainfall per day. Older children could consider astronomical phenomena such as the positions of the sun and moon, or measurements made on an electronic circuit, or observations of a pendulum, or price movements of two different exchange rates on the world currency market.

The oscillations of a simple pendulum provide a very illuminating example of a phase portrait—but suitable in full detail only for more advanced students. The traditional approach to the pendulum is to write down an approximate equation whose solution is a sine curve. The approximation is necessary because standard techniques of calculus cannot solve the true equation for an exact model. The student does learn useful properties of the sine curve as well as a formula for the period of a pendulum whose swings are small. However, this traditional approach is in some respects unsatisfactory since the approximations employed are rarely justified. It leaves the unwarranted impression that lack of precision is acceptable in mathematics.

Instead, the law of conservation of energy can be applied to yield an exact model for the motion of a pendulum. It leads to the equation

$$v = C\sqrt{k + 2\cos\theta},$$

where v is velocity, C and k are constants, and θ is the angle that the pendulum makes with the vertical. By sketching this family of curves, one in effect draws the phase portrait (Figure 8). All of the motions of a real pendulum, including large swings—even cases when it revolves like a propeller—can be seen in this picture.[24] With this alternate approach, students obtain equally valid practice with the sine function, an accurate

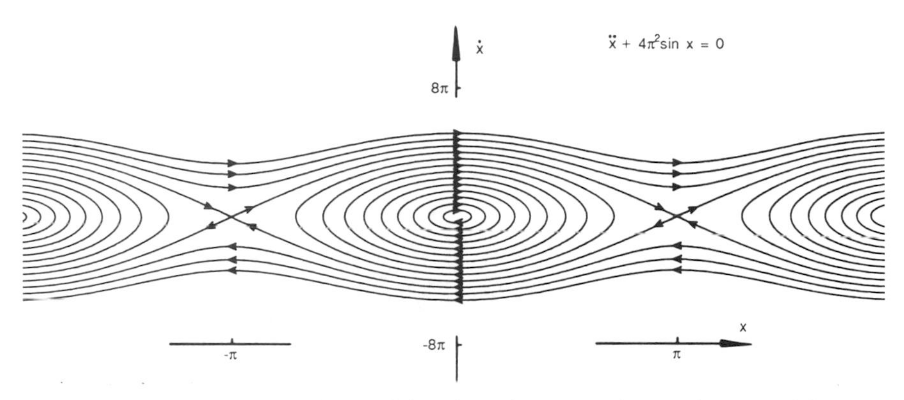

FIGURE 8. Phase portrait of a pendulum in which all possible motions are visible.

model, no approximations, and an important physical principle (conservation of energy). Isn't that a better way to think about the pendulum?

Resonance

The dynamical equations for three bodies cannot be solved by a formula, but they can be put on a computer and solved numerically. Such models provide a good means of exploring the surprising effects of resonance on the motion of dynamical systems. Resonance occurs when different periodic motions have periods that are in some simple numerical relationship such as 1:1, 2:1, 3:2, and so on. For example, Titan, a satellite of Saturn, has an orbital period that is close to 4:3 resonance with that of another satellite, Hyperion. Specifically, Hyperion takes 21.26 days to complete one orbit and Titan takes 15.94. The ratio of these is 1.3337, convincingly close to the ratio 4:3.

Older children can use a computer package to simulate planetary dynamics. They can study the motion of the moon or of a satellite in transit from earth to moon. They can study the way in which Jupiter's satellites are locked into resonant orbits. They can study the so-called Lagrange points, where satellites (or space colonies) can remain in stable positions 60° ahead of or behind the moon. This too is a kind of resonance.

Resonances are especially important in dynamics. They lead to a rich and subtle geometry that is almost unbelievably complex. In Figure 9 the large circles represent regular motion; secondary "islands" between the circles represent resonances; tertiary islands signal more delicate multiple resonances. The spaghetti-like crossings represent chaos. The structure repeats forever on smaller and smaller scales.

FIGURE 9. Fractal structure near a periodic orbit: islands signal resonances of various orders, while tangles represent regions of chaos.

High school students can easily search astronomical tables to look for evidence of resonances. This work involves plenty of practice with fractions, decimals, calculators, and computers. It shows how simple mathematics can produce deep insights to those who look at the world from a mathematical perspective.

Resonances often generate chaos. Figure 9 has a particular disturbing quality of *self-similarity:* each island has the same complexity, indeed the same qualitative form, as the entire picture. This complicated self-similar structure is not some mad mathematician's nightmare. It's what really happens.

The concept of self-similarity together with the associated ideas of fractal geometry,[14] can be made accessible to children around the age of twelve, maybe younger. The topic can be introduced using natural examples: coastlines, leaves, ferns, etc. Next, computer models of fractals such as the Cantor set and snowflake and dragon curves can be drawn and their patterns analyzed. Concepts of fractal structure and self-similarity can easily be developed from these examples. Even young children can appreciate the idea of fractal dimensions—which need not be whole numbers.

Gaps and Clumps

Resonances feature prominently in another astronomical conundrum, the gaps in the asteroid belt, which is directly related to our original question about meteorites. Most asteroids circle between the orbits of Mars and Jupiter, although a few come much closer to the sun. However, the asteroid orbits are not spread uniformly between Mars and Jupiter. Their radii tend to cluster around some values and stay away from others (Figure 10). Daniel Kirkwood, an American astronomer who called attention to this lack of uniformity in about 1860, also noticed an intriguing feature of the most prominent gaps: if an asteroid were to orbit the sun in one of these *Kirkwood gaps,* then its orbital period would resonate with that of Jupiter. Conclusion: Resonance with Jupiter somehow perturbs any bodies in such orbits, causing some kind of instability that sweeps them away to distances at which resonance no longer occurs. The special role of Jupiter is no surprise since it is so massive in comparison with the other planets. The gaps are obvious in recent data, especially at resonances 2:1, 3:1, 4:1, 5:2, and 7:2. On the other hand, at the 3:2 resonance there is a *clump* of asteroids, the Hilda group. So stability is not *just* a matter of resonance: it depends on the type of resonance. The questions remain a subject of intense investigation.

Recent computer calculations[30] show that an asteroid orbiting at a distance that would suffer 3:1 resonance with Jupiter can either follow a roughly circular path or a much longer and thinner elliptical path. If the orbit of an asteroid is sufficiently elongated, it crosses the orbit of Mars. Every time it does so there is a chance that the asteroid will come sufficiently close to Mars for its orbit to be severely perturbed. It will eventually come too close and be sent off into some totally different orbit. The 3:1 Kirkwood gap is there because Mars sweeps it clean, rather than being due to some action of Jupiter. What Jupiter does is create the resonance that causes the asteroid to become a Mars crosser; then Mars kicks it away into the cold and dark. Jupiter creates the opening; Mars scores.

The same mechanism that causes asteroids to be swept up by Mars can also cause meteorites to reach the orbit of the earth. The 3:1 resonance

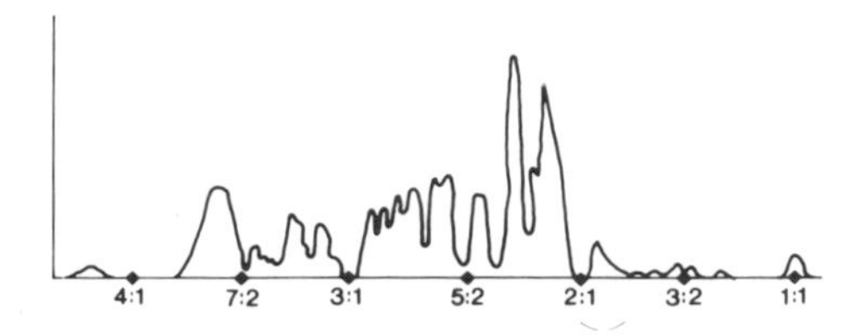

FIGURE 10. Gaps and clumps in the distribution of asteroids reveal resonance with the orbital period of Jupiter.

with Jupiter thus appears to be responsible for transporting meteorites from the asteroid belt into earth orbit, to burn up in our planet's atmosphere if they hit it.[26] A cosmic football game, played among the asteroids by Mars and Jupiter, determines whether or not floating cosmic rocks—and perhaps sometimes mountains—will crash into the earth's atmosphere. It would be hard to find a more dramatic example of the essential unity of the entire solar system or a better example of the interconnectedness of change.

THE TIGER'S STRIPES

"What immortal hand or eye dare frame thy fearful symmetry?" said William Blake, referring to the tiger. Although Blake wasn't using the word "symmetry" in a technical sense, it turns out that the behavior of symmetric systems has a distinct bearing on the striped nature of tigers.

Symmetry is basic to our scientific understanding of the universe.[13] The symmetries of crystals not only classify their shapes but also determine many of their properties. Many natural forms—from starfish to raindrops, from viruses to galaxies—have striking symmetries. Man-made objects also tend to be symmetric: cylindrical pipes, circular plates, square boxes, spherical bowls, hexagonal steel bars.

That symmetric causes have symmetric effects is a long-standard principle in the folklore of mathematical physics. Pierre Curie made the case succinctly:[6] "If certain causes produce certain effects, then the symmetries of the causes reappear in the effects produced." The principle seems natural enough—but is it true? The question is a subtle one involving not just the meaning of "symmetry" but also that of "cause" and "effect."

Recently scientists and mathematicians have become aware that, in an important sense, Curie's statement is false. It is possible for a symmetric system to behave in an asymmetric fashion. This phenomenon, known as *symmetry breaking,* is an important mechanism underlying pattern formation in many physical systems from astronomy to zoology. The mathematical theory of symmetry breaking provides a powerful method for analyzing how symmetric systems behave and applies across the entire range of scientific disciplines.[12]

Curie Was Right ...

At first glance, Curie's statement is "obviously" true. If a planet in the shape of a perfect sphere acquires an ocean, that ocean will surely be of uniform depth, hence itself a sphere. The spherical symmetry of the planet is reflected in a corresponding spherical symmetry of its ocean.

It would appear bizarre if, in the absence of any asymmetric cause, the ocean should decide to bulge unevenly.

On the other hand, if the planet rotates—breaking the spherical symmetry and replacing it by circular symmetry about the axis of rotation—then the ocean will bulge at the equator, preserving the circular symmetry. Isn't that typical of how symmetry behaves? Not always.

Curie Was Wrong ...

Curie's principles may seem obvious, but they must be interpreted *very* carefully indeed, for there are many symmetric systems whose behaviors are less symmetric than the full system. For example, if a perfect cylinder, say a tubular metal strut, is compressed by a sufficiently large force, it will buckle.[28] The buckling is *not* a consequence of lack of symmetry caused by the force: even if the force is directed perfectly along the axis of the tube, preserving the rotational symmetry about that axis, the tube will still buckle. Buckled cylinders cease to be cylindrical—that's what "buckle" means. Similarly, a computer picture of a spherical shell buckled by a spherically symmetric compressive force is shown in Figure 11: observe that the symmetry of the buckled state is *circular* rather than spherical.

It is important to understand that the loss of symmetry in these systems is *not* merely a consequence of small imperfections: asymmetric solutions will exist even in an idealized perfectly symmetric mathematical system. Indeed, such a "perfect" system largely controls *how* symmetries can break. However, imperfections play an important role in selecting exactly *where.* For example, when a perfect system such as the sphere in Figure 11 buckles, the axis of circular symmetry can be

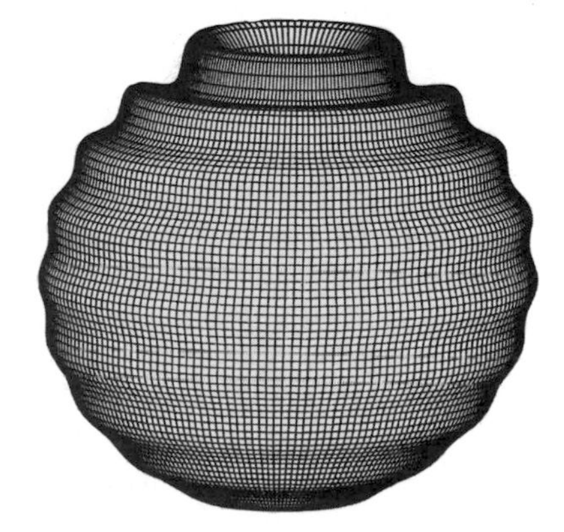

FIGURE 11. Symmetry-breaking buckling of a uniform spherical shell subjected to uniform external pressure. The shell buckles in a cylindrically symmetric fashion.

any axis of the original sphere; for an imperfect system some axes will be preferred, their positions being related to weaknesses in the spherical shell. The general form of the buckled sphere, however, will be the same in both cases.

In this sense Curie's principles are perhaps valid for an actual physical system (which is necessarily imperfect) but not for an idealized model. Rather than attempting to resurrect Curie's principles in this fashion, however, it seems preferable to understand the mechanism by which perfect idealized symmetric systems produce behavior with less symmetry. This is called *symmetry breaking.* It seems to be responsible for many types of pattern formation in nature, and it has a very well defined mathematical structure that can be used to understand such processes.

What causes the symmetry to break? The answer is that natural systems must be *stable.* Curie was right in asserting that symmetric systems should have symmetric states, but he failed to address their stability. If a symmetric state becomes unstable, then the system will do something else—and that something else cannot be symmetric.

How does the symmetry "get lost"? We answer this question by an example. The catastrophe machine (Figure 12), invented for rather different reasons by Christopher Zeman of Warwick University in 1969,[19,20,31] shows that symmetry is not so much broken as spread around. Children can make one and experiment with it.

The entire catastrophe machine has reflectional symmetry about the center line. If you begin to stretch the free elastic, the system obey's Curie's principles and stays symmetric; that is, the disk does not rotate (Figure 13a). But as you stretch the elastic further, the disk suddenly begins to turn—maybe clockwise, maybe counterclockwise (Figure 13b). Now the state of the system loses its reflectional symmetry. The symmetry has broken, and Curie's principles have failed.

Where has the missing symmetry gone? Hold the elastic steady and rotate the disk to the symmetrically placed position on the other side (Figure 13c). You will find that it remains there. Instead of a single symmetric state we have two *symmetrically related* states.

This is a general feature of symmetry breaking. The system can exist in several states, each obtainable from the others by one of the symmetries of the full system. For example, the buckled spherical shell in Figure 11 breaks symmetry from spherical to circular, and the circular symmetry occurs about some particular axis, clearly visible in the picture. In the "perfect" system any axis is possible, but all buckled states have the identical shape, and they differ only by motions of the sphere.

Children can explore symmetry breaking with simple experiments. They can compress a plastic ruler to find out when and how it bends.

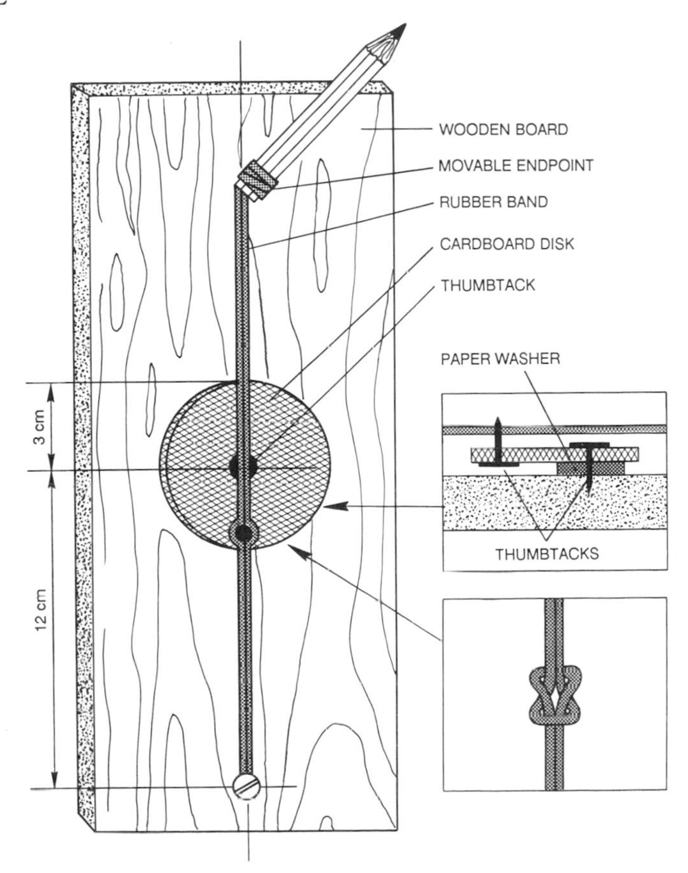

FIGURE 12. A "catastrophe machine" can be constructed easily out of cardboard and rubberbands. Attach a circular disk of thick cardboard, of radius 3 centimeters, to a board using a drawing pin and a paper washer. Fix another drawing pin near the rim of the disk with its point upwards. To this pin attach two elastic bands, of about 6 centimeters unstretched length. Fix one to a point 12 centimeters from the center of the disk, and leave the end of the other free to move along the center line as shown, for example, by taping it to a pencil that you can move by hand.

They can use a spring to hold a rod upright, with the lower end resting on a table, then add weights to the top and watch it sway or buckle. They can make a "bridge" from a flexible metal strip, put weights on top, and watch it collapse.

Older students can analyze the behavior of two rigid rods joined by a springy hinge. These models lead naturally to more subtle questions relating symmetry, stability, and continuous change. How does a rolling body move if its center of gravity changes? How do ships capsize? The analysis of models of such changes brings in a great deal of important

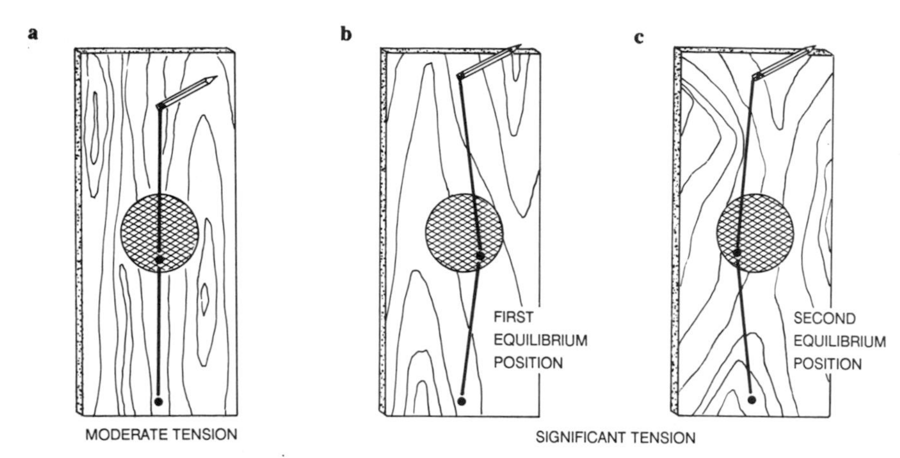

FIGURE 13. When the rubber band is stretched, the symmetrical position of the pin (a) becomes unstable. Two stable positions emerge on either side (b) and (c), but neither of these has the symmetry of the original configuration. In this case, as in many other examples in nature, instability break symmetry.

geometry, for example, tangents and normals to a curve, centers of gravity, and even coordinate transformations.

For a more homely example, consider the flow of water through a hose with circular cross section. Imagine the hose suspended vertically, nozzle downwards, with water flowing steadily through it. This system is circularly symmetric about an axis running vertically along the center of the hose. And indeed if the speed of the water is slow enough, the hose just remains in this vertical position, retaining its circular symmetry.

However, if the faucet is turned on further, the hose will begin to wobble. In fact there are two distinct kinds of wobble. In one it swings from side to side like a pendulum. In the other it goes round and round, spraying water in a spiral. Similar effects are often observed when children wash the family car. These wobbles do not possess circular symmetry about a vertical axis: indeed, they break it in two distinct ways. They also break a less obvious but very important symmetry: symmetry in time. The original steady flow looks exactly the same at all instants of time. The oscillating flows do not. The time symmetry is not totally lost, however: both wobbles are periodic and hence look exactly the same when viewed at times that are whole number multiples of the period. This shows how the continuous temporal symmetry of a steady state breaks to give the discrete symmetry of a periodic one.

Symmetry breaking is important in biology. When a spherically symmetric frog egg develops, it splits into two cells and the spherical symmetry is broken. At a later stage of development (Figure 14) a spherically symmetric mass of cells, the *blastula,* forms; but this first develops

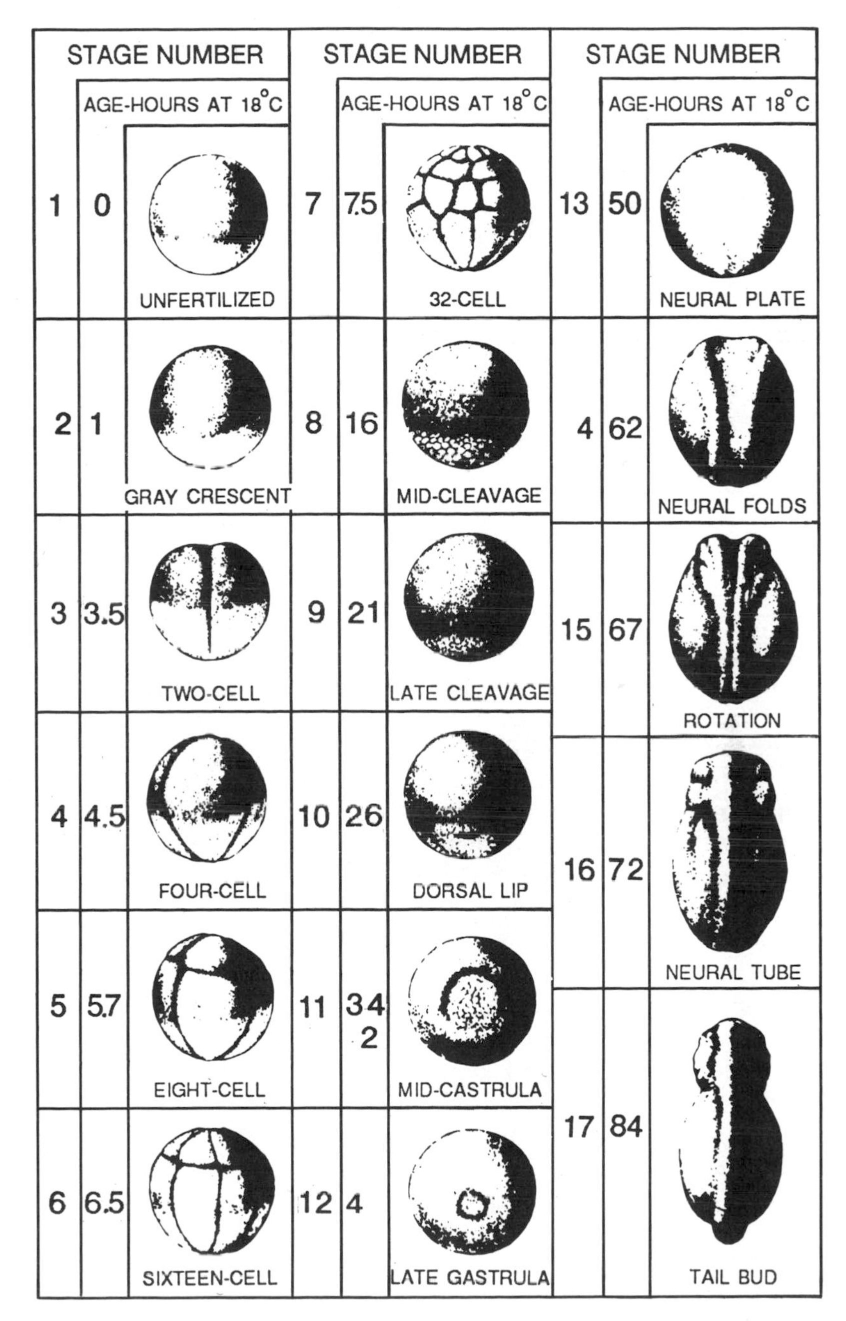

FIGURE 14. Creation and destruction of symmetry in the development of a frog embryo: spherical symmetry breaks, then is restored, then breaks again into circular and then bilateral symmetry.

a circular dent (gastrulation) with only circular symmetry and then a neural fold, leading to mere bilateral symmetry.

Mathematically, the development of a circular dent during gastrulation is directly analogous to the buckling of a spherical shell (Figure 11). This demonstrates that symmetry-breaking phenomena in quite different physical realizations can have the same underlying mathematical structure. The unifying role of mathematics in science, one of its most striking and important features, is clearly visible.

This leads directly to our motivating question: Why does a tiger, with its roughly cylindrical symmetry, have *stripes?* Blake's immortal poem offers no useful clue.

Turing's Tiger

The theory behind the tiger's stripes goes back to Alan Turing, more famous as one of the father figures of modern computing. Turing knew that chemical changes produce the variations in coloring. The chemical responsible for the stripes need not be the actual coloring matter; it is more likely to be a precursor, formed during relatively early stages of the tiger's development, which later triggers a series of chemical changes to create the stripes. However, the biological details—some of which remain controversial—are not important here. Our aim is to illustrate some simple and general mathematical mechanisms for pattern formation in a familiar context.

Turing wrote down equations for this kind of chemical change.[29] He solved them numerically and then made pictures of the results. He used to button-hole friends and show them his pictures. On some there were stripes, on others irregular patches. "Don't these look just like the markings on cows?" Turing would ask, in some excitement.

His calculations showed that patterns like stripes or spots can be created by a mechanism of instability. Imagine a flat surface (mathematical tiger skin) that contains a uniform distribution of some chemical. This would in the course of time produce a tiger of uniform color, grayish brown all over, more like a mountain lion. But the distribution of chemicals need not remain uniform: it can change. There are two important types of change. Chemicals at a given place react, and the reaction products diffuse from one place to another.

These two types of change compete. Reaction tries to alter the chemical mix; diffusion tries to make it the same everywhere. The mathematics shows that when different influences compete the result is often a compromise. Here the simplest compromise is that the uniform distribution of chemicals begins to form ripples. If instability occurs in only

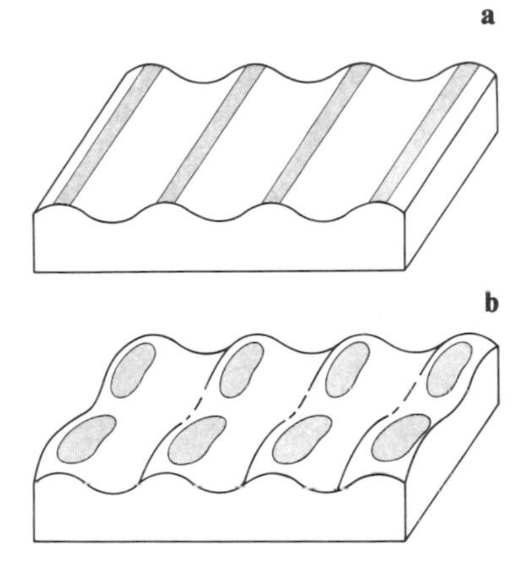

FIGURE 15. Competing chemical forces lead to instabilities. In (a), instability in one direction leads to stripes; in (b), instability in a second direction breaks up the stripes into spots.

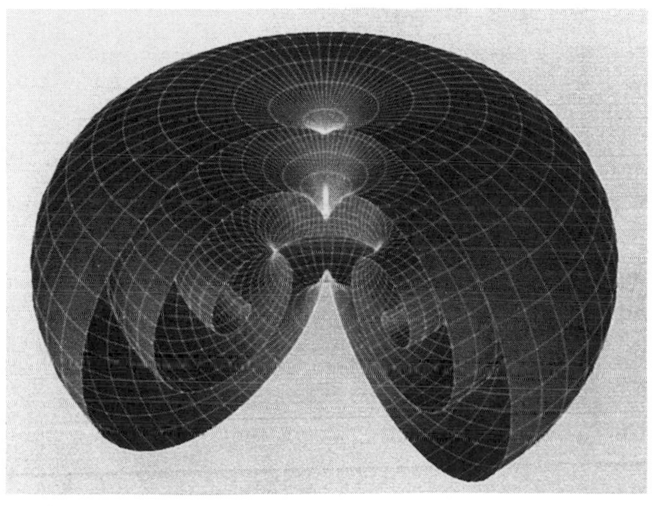

FIGURE 16. Spiral scroll waves in a chemical reaction created by conflicting roles of reaction (which changes the chemical mix) and diffusion (which restores uniformity).

one direction, then the ripples only run one way and we see stripes. If a second instability sets in along a perpendicular direction, then the stripes themselves ripple along their length and break up into spots (Figure 15). Competing chemical instabilities may well be the fundamental difference, on a mathematical level, between tigers and leopards.

Chemical reactions that can generate periodic patterns—spirals, target patterns—can be demonstrated in any school chemistry laboratory (Figure 16). The most famous one is the so-called Belousov-Zhabotinskii reaction.[21] Students can analyze these patterns to find their mathematical structure (e.g., what sort of spiral is it?). They can also use computer

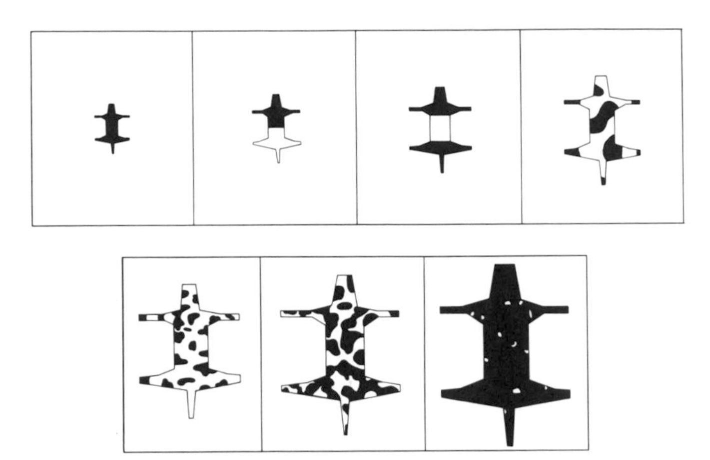

FIGURE 17. By using simple computer packages, children can explore reaction-diffusion in regions of different shapes.

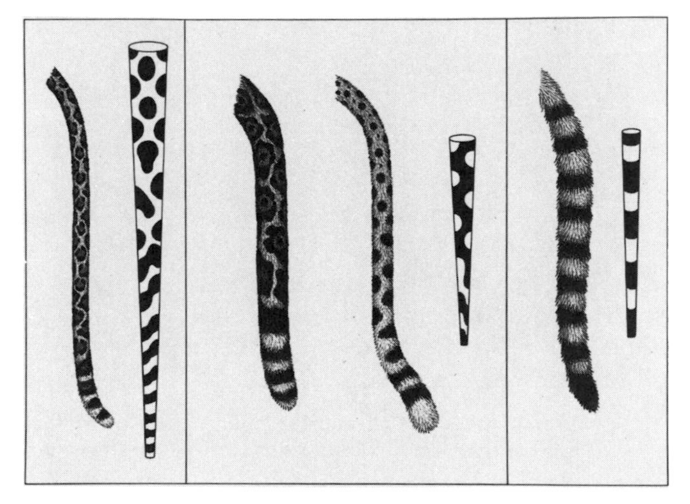

FIGURE 18. Computer models of patterns on animal skins show realistic-looking results. They also show that long thin stripes usually break up into spots.

packages to solve reaction-diffusion equations on different shapes of regions and see what kinds of patterns occur (Figure 17).

Patterns formed in the competition between reaction and diffusion provide good examples of symmetry breaking. The initial uniform distribution of chemicals has greater symmetry than do the stripes or spots or spirals. Symmetry breaking is a very common source of natural patterns. And what else is the breaking of symmetry than a change in pattern?

Computer models of how pigmentation-controlling chemicals might diffuse through the tiger's tail produce plausible markings (Figure 18). Long thin stripes are less stable than short fat ones and prefer to break up into spots.[16] This mathematical result could help explain the common observation that a spotted animal can have a striped tail, but a striped animal cannot have a spotted tail.

IMPLICATIONS

Change is a phenomenon that has a direct impact on every human being. It affects individual lives, national economies, and the future of the entire planet. Until recently our understanding of change came mostly from the traditional tools of calculus and its more advanced relatives and was confined to the physical sciences, where accurate numerical measurements are possible.

Initially, computers served to extend the techniques of calculus, by making it possible to solve more difficult equations. The term "number crunching" captures the style. But today's computers do more than just crunch numbers. In particular they can represent and manipulate data graphically. As a complementary development, today's mathematics is also about far more than just numbers. It deals in structural features, multidimensional spaces, transformations, shapes, forms—in short, *patterns.*

When calculus was invented, it evolved hand in hand with geometry. Over the centuries, geometric reasoning was replaced by more powerful—but less informative—analytic techniques. The emphasis shifted to formulas. Now, as we penetrate areas where formulas alone are inadequate, the emphasis is shifting back to geometry—not to the stilted formal reasoning often associated with the school treatment of geometry, but to the geometry of space and shape—to the mathematics of the visual.

Many basic skills are involved, often as complementary pairs, to provide two different ways to approach the same problems:

- numerical and visual,
- algebraic and geometric,
- formal and experimental,
- abstract and concrete,
- analytic and synthetic,
- algorithmic and existential,
- conceptual and computational.

Mathematics, the science of patterns, is itself changing. For the sake of our future we must harness mathematics to the patterns of change. And to do that we must change the way that mathematics is taught, to create a new generation able to perceive and manipulate new patterns.

REFERENCES AND RECOMMENDED READING

1. Abraham, Ralph and Shaw, Christopher D. *Dynamics: The Geometry of Behavior, Volumes 1–4.* Santa Cruz, CA: Aerial Press, 1983.
2. Arnold, V.I. *Catastrophe Theory.* New York, NY: Springer-Verlag, 1984.
3. Becker, Karl-Heinz and Dörfler, Michael. *Dynamical Systems and Fractals: Computer Graphics Experiments in Pascal.* Cambridge, MA: Cambridge University Press, 1989.
4. Beltrami, Edward. *Mathematics for Dynamic Modelling.* Boston, MA: Academic Press, 1987.
5. Crutchfield, James P.; Farmer, J. Doyne; Packard, Norman H.; Shaw, Robert S. "Chaos." *Scientific American,* (December 1986), 38–49.
6. Curie, M.P. "Sur la symétrie dans les phénomènes physiques, symétrie d'un champ électrique et d'un champ magnétique." *Journal de Physique,* 3, Series 3, (1894) 393–415.
7. Devaney, Robert L. *An Introduction to Chaotic Dynamical Systems.* Menlo Park, CA: Benjamin-Cummings, 1986.
8. Devlin, Keith. *Mathematics: The New Golden Age.* London, England: Penguin, 1988.
9. Ekeland, Ivar. *Mathematics and the Unexpected.* Chicago, IL: University of Chicago Press, 1988.
10. Garfunkel, Solomon and Steen, Lynn A. (Eds.). *For All Practical Purposes.* New York, NY: W.H. Freeman, 1988.
11. Gleick, James. *Chaos: Making a New Science.* New York, NY: Viking Press, 1987.
12. Golubitsky, Martin; Stewart, Ian; Schaeffer, David G. *Singularities and Groups in Bifurcation Theory, Volume 2.* New York, NY: Springer-Verlag, 1988.
13. Hargittai, István and Hargittai, Magdolna. *Symmetry Through the Eyes of a Chemist.* Weinheim, FRG: VCH Publishers, 1986.
14. Mandelbrot, Benoît. *The Fractal Geometry of Nature.* San Francisco, CA: W.H. Freeman, 1982.
15. May, Robert M. "Mathematical aspects of the dynamics of animal populations." In Levin, S.A. (Ed.): *Studies in Mathematical Biology.* Washington, DC: Mathematical Association of America, 1978.
16. Murray, James D. "How the leopard gets its spots." *Scientific American,* 258 (March, 1988), 62–69.
17. Peitgen, Heinz-Otto and Richter, Peter H. *The Beauty of Fractals.* New York, NY: Springer-Verlag, 1986.
18. Peterson, Ivars. *The Mathematical Tourist.* New York, NY: Freeman, 1988.
19. Poston, Tim and Stewart, Ian. *Catastrophe Theory and Its Applications.* Boston, MA: Pitman, 1978.
20. Poston, Tim and Woodcock, A.E.R. "On Zeeman's catastrophe machine." *Proceedings of the Cambridge Philosophical Society,* 74 (1973), 217–226.
21. Prigogine, Ilya. *From Being to Becoming.* San Francisco, CA: Freeman, 1980.
22. Stewart, Ian. *The Problems of Mathematics.* Oxford, England: Oxford University Press, 1987.

23. Stewart, Ian. "The nature of stability." *Speculations in Science and Technology,* 10 (1988), 310–324.
24. Stewart, Ian. *Does God Play Dice? The Mathematics of Chaos.* Oxford, England: Blackwell, 1989.
25. Stewart, Ian. "Chaos: Does God Play Dice?" *1990 Yearbook of Science and the Future.* Chicago, IL: Encyclopaedia Britannica, 1989, 54–73.
26. Stewart, Ian. "Dicing with death in the solar system." *Analog,* 109 (1989), 57–73.
27. Thompson, D'Arcy. *On Growth and Form, Volumes 1 & 2.* Cambridge, MA: Cambridge University Press, 1942.
28. Thompson, J.M.T. and Stewart, H.B. *Nonlinear Dynamics and Chaos.* New York, NY: John Wiley & Sons, 1986.
29. Turing, A.M. "The chemical basis of morphogenesis." *Philosophical Transactions of the Royal Society of London,* 237, Series B (1952), 37–72.
30. Wisdom, J. "Chaotic behaviour in the solar system." In Berry, M.V.; Percival, I.C.; Weiss, N.O. (Eds.): *Dynamical Chaos.* London, England: The Royal Society, 1987, 109–129.
31. Zeeman, E.C. "A catastrophe machine." In Waddington, C.H. (Ed.): *Towards a Theoretical Biology, Volume 4.* Edinburgh, England: Edinburgh University Press, 1972, 276–282.

Biographies

THOMAS BANCHOFF has been a professor in the mathematics department at Brown University for 23 years. A 1960 graduate of the University of Notre Dame, Banchoff received his Ph.D. at the University of California, Berkeley in 1964. He has held a Fulbright at the University of Amsterdam and served as Benjamin Peirce Instructor at Harvard University. Banchoff is the author of approximately fifty articles and monographs, primarily dealing with geometry, many illustrated by computer graphics images. His film with Charles Strauss, "The Hypercube: Projections and Slicing," has received international awards. Banchoff's most recent book *Beyond the Third Dimension: Geometry, Computer Graphics, and Higher Dimensions* was issued as part of the Scientific American Library. He was awarded a Lester Ford Award for exposition in mathematics and the Joseph Priestley Medal from Dickinson College in 1988.

JAMES FEY is professor of curriculum and instruction and mathematics at the University of Maryland in College Park, where he has been since 1969. His bachelors and masters degrees were earned in mathematics and mathematics education, respectively, from the University of Wisconsin; in 1968 he received a doctorate in mathematics education from Teachers College, Columbia University. At Maryland Fey teaches mathematics content and methods courses aimed primarily at prospective and in-service teachers of secondary mathematics. His main research interest is the development of innovative mathematics curricula. For the past ten years that work has focused on curricula that use calculators and computers as learning and problem solving tools. Fey is issue editor for the 1992 NCTM Yearbook on calculators.

DAVID MOORE is professor of statistics at Purdue University, where he has been a member of the faculty since 1967. He received his A.B. from Princeton University and the Ph.D. from Cornell University. Moore has written many research papers and several books, has served on the editorial boards of leading journals, and has served as statistics program director for the National Science Foundation. He is a fellow of the American Statistical Association and of the Institute of Mathematical Statistics and is a member of the International Statistical Institute. Moore has a long-standing interest in statistical education; he was, for example, the content developer for the Corporation for Public Broadcasting's telecourse "Against All Odds: Inside Statistics," and is author or co-author of two widely used elementary textbooks, *Introduction to the Practice of Statistics* and *Statistics: Concepts and Controversies.*

MARJORIE SENECHAL received her Ph.D. in 1965 from the Illinois Institute of Technology. Since 1966 she has taught at Smith College, where she is Louise Wolff Kahn Professor of Mathematics. Her field of research is mathematical crystallography, a broadly interdisciplinary subject focusing on the classification of geometrical patterns in the plane and in space. Senechal is the author of *Crystalline Symmetries: An Informal Mathematical Introduction*, a monograph on crystallography for physicists, mathematicians, and metallurgists. Other professional interests include the history of science and geometry education. Together with the chemist George Fleck, she was an organizer of "The World's First Symmetry Festival" in 1973 and was co-editor of *Patterns of Symmetry*, a volume of essays based on it; in 1984 they organized an interdisciplinary geometry festival which led to the volume *Shaping Space: A Polyhedral Approach.*

LYNN ARTHUR STEEN is professor of mathematics at St. Olaf College in Northfield, Minnesota. He received his Ph.D. in 1965 from the Massachusetts Institute of Technology, and his B.A. from Luther College in Iowa. Steen is the editor or author of ten books, including *Everybody Counts*, *Calculus for a New Century*, *Mathematics Today*, and *Counterexamples in Topology*. He has written numerous articles about mathematics, computer science, and mathematics education for periodicals such as *Educational Leadership, Daedalus, Scientific American, Science News*, and *Science.* Steen is Co-Director of the Minnesota

Mathematics Mobilization, Telegraphic Reviews Editor for the *American Mathematical Monthly*, and chair of the Committee on the Undergraduate Program in Mathematics (CUPM). In previous years, he has served as President of the Mathematical Association of America, Secretary of Section A (Mathematics) of the American Association for the Advancement of Science, and Chairman of the Conference Board of the Mathematical Sciences.

IAN STEWART was born in Folkestone, England, in 1945. He was educated at the University of Cambridge, obtaining a B.A. degree in mathematics in 1966, and at the University of Warwick, where he gained a Ph.D. in 1969. He has held positions at the universities of Tübingen, Auckland, Connecticut, and Houston and is currently reader in mathematics at Warwick. His research area is nonlinear systems and bifurcation theory. He has written a large number of books, including *Concepts of Modern Mathematics, The Problems of Mathematics, Does God Play Dice?,* and *Game, Set, and Math.* He is European editor of the *Mathematical Intelligencer,* and has written articles on mathematics for *Scientific American, New Scientist,* and *The Sciences.* His column "Visions Mathematiques" appears regularly in the French, German, Italian, Spanish, and Japanese editions of *Scientific American.* He has worked in television and makes brief but regular appearances on BBC radio.

Index

~ ~ ~ ~ ~ ~ ~ ~ ~ ~ ~ ~ ~ ~ ~ ~ ~ ~ ~ ~

Credits for

ON THE SHOULDERS OF GIANTS
New Approaches To Numeracy

MATHEMATICAL SCIENCES EDUCATION BOARD

Executive Director, Kenneth M. Hoffman
Staff Coordination, Linda P. Rosen
Senior Project Assistant, Jana Godsey
Senior Project Assistant, Carol Metcalf

NATIONAL ACADEMY PRESS

Editorial Coordination, Sally Stanfield
Marketing Coordination, Barbara Kline
Production Coordination, Dawn M. Eichenlaub
Cover Design, Francesca Moghari
Graphics, James Butler

COMPOSITION COURTESY OF

American Mathematical Society

FINANCIAL SUPPORT

We wish to thank the Carnegie Corporation of New York for support of the development, publication, and dissemination of this document. We also wish to thank the Andrew W. Mellon Foundation for their support of further dissemination.

PHOTOGRAPHS COURTESY OF

Texas Instruments, Inc. (p. 63)
Stan Sherer (pp. 148, 152, 178)
Marjorie Senechal (p. 149)

COMPUTER ARTWORK COURTESY OF

Thomas Banchoff
Davide Cervone
David Moore

ARTWORK COURTESY OF

J. Weeks, *The Shape of Space*, Marcel Dekker, Inc. (pp. 143, 144)

P. Stevens, *Patterns in Nature*, Little, Brown & Company (p. 147)
Color Treasury of Crystals, Crescent Books (p. 152)

R.J. Hauy, *Cristallographie, Cultures et Civilisations* (p. 156)

B. Grünbaum and G.S. Shephard, *Tilings and Patterns*, W.H. Freeman (pp. 157, 159)

N.J.W. Thrower, *Maps and Man*, Prentice-Hall (p. 162)

D. Bruyr, *Geometrical Models & Demonstrations*, J. Weston Walch (p. 164)

"Draftsman Drawing A Reclining Nude," by Albrecht Durer in *Complete Engravings, Etchings & Woodcuts*, Courtesy of the Library of Congress (p. 165)

Mathematical Association of America (p. 165)

M.C. Escher Heirs, Cordon Art—Baarn, Holland (p. 166)

B. Grünbaum, "Geometry Strikes Again," *Mathematics Magazine*, Mathematical Association of America (p. 167)

Thomas Harriot's lunar drawing, July 26, 1609, Petworth Museum (p. 169)

Galileo's lunar drawing, Biblioteca Nazionale (Florence, Italy) (p. 169)

IBM-Almaden Research Center (p. 170)

W.A. Bentley & W.J. Humphries, *Snow Crystals*, Dover Publications (p. 174)

R. Abraham and J. Marsden, *Foundations of Mechanics*, Addison-Wesley Publishing Co., Inc. (p. 185)

R. Abraham and C.D. Shaw, *Dynamics: The Geometry of Behavior*, Aerial Press (p. 185)

I. Stewart, *Does God Play Dice?* Basil Blackwell, Inc. (pp. 185, 199, 201, 205)

J.M.T. Thompson & H.B. Stewart, *Nonlinear Dynamics and Chaos*, John Wiley & Sons Ltd. (pp. 185, 203)

G. Oster in S.A. Levin (ed.) *Studies in Mathematical Biology*, Mathematical Association of America (p. 195)

T. Poston, Pohang Institute of Science & Technology (p. 204)

M. Shumway, "Stages in the Normal Development of Rana Pipiens I, II," *Anatomical Record*, Alan R. Liss, Inc. (p. 211)

J.M.T. Thompson, *Instabilities and Catastrophes in Science and Engineering*, John Wiley & Sons Ltd. (p. 211)

A. Winfree and S. Strogatz, *Mathematical Intelligencer*, Volume 7, No. 2 (cover), Springer-Verlag Publishers (p. 213)

J.D. Murray, "How the Leopard Gets Its Spots," *Scientific American.* Illustration by Patricia J. Wynne (p. 214)